Macromolecular Symposia

Editor-in-chief: Hartwig Höcker
Editors: W. Guth, B. Jung, I. Meisel, S. Spiegel

Symposium Editors: H.-J. P. Adler, D. Kuckling, J. C. Jung

[redacted]

142

WILEY-VCH

pp. 1–225

August 1999

Macromolecular Symposia publishes lectures given at international symposia and is issued irregularly, with normally 12 volumes published per year. For each symposium, the Symposium Editor(s) is/are appointed. The journal is produced by photo-offset lithography directly from the authors' typescripts.
Further information for authors can be obtained from
Wiley-VCH Verlag GmbH
Editorial office "Macromolecular Chemistry and Physics"
P. O. Box 10 11 61, D-69451 Weinheim, Germany
Tel. +49 (0) 62 01/6 06-2 34 or -2 38; Fax +49 (0) 62 01/6 06-5 10; macromol@wiley-vch.de
Authors bear the entire responsibility for their work and must be sole owners of the copyright. Acceptance of a manuscript by the editors implies the transfer to the publishers of the exclusive rights of use. Requests for permission to reproduce without payment parts of articles or complete articles for scientific purposes should be sent to the editorial office in Mainz at the address given above.

More information on our journals, including the "Notes for Contributors to *Macromol. Symp.*", can be found at the following WWW address:

http://www.wiley-vch.de/home/macrosymp

Annual subscription rates 2000
Germany, Austria € 1018 (DM 1991,03); Switzerland SFr 1678; other Europe € 1018; outside Europe US $ 1228.
Macromolecular Basic Package, including Macromolecular Chemistry & Physics (18 issues), Macromolecular Rapid Communications (18 issues), Macromolecular Theory & Simulations (9 issues): Germany, Austria € 2928 (DM 5726,67); Switzerland SFr 4818; other Europe € 2928; outside Europe US $ 3578.
Combined subscription packages with Macromolecular Symposia and Macromolecular Materials & Engineering also available. Please ask for details at subservice@wiley-vch.de

Single issues and back copies are available. Please, inquire for prices.

Orders may be placed through your bookseller or directly at the publishers: WILEY-VCH Verlag GmbH, P. O. Box 10 11 61, D-69451 Weinheim, Tel.: (0 62 01) 6 06–0, Telefax (0 62 01) 60 61 17, Telex 46 55 16 vchwh d.
E-mail: subservice@wiley-vch.de

Macromolecular Symposia (ISSN 1022-1360) is published with 12 volumes per year by WILEY-VCH Verlag GmbH, P.O. Box 10 11 61, D-69451 Weinheim, Germany. Air freight and mailing in the USA by Publications Expediting Inc., 200 Meacham Ave., Elmont, NY 11003.
Periodicals postage pending at Jamaica, NY 11431.
POSTMASTER: send address changes to Macromolecular Symposia, Publications Expediting Inc., 200 Meacham Ave., Elmont, NY 11003.

Printing: Strauss Offsetdruck, Mörlenbach
Binding: Wilh. Osswald, Neustadt

4th German-Korean Symposium Advanced Polymeric Materials – New Synthetic Routes and Applications

held in Dresden, Germany
September 29–30, 1998

Symposium Editors

Hans-Juergen P. Adler, Dirk Kuckling
Institute of Macromolecular Chemistry and Textile Chemistry
Dresden University of Technology
D-01062 Dresden, Germany

Jin Chul Jung
Department of Materials Science and Engineering
Pohang University of Science and Technology
San 31, Hyoja-dong, Pohang, 790-784 Korea

ISBN 3–527–29902–5

The **tables of contents** of the published issues are displayed on the www.

This service as well as further information on our journals can be found at the following www address:

http://www.wiley-vch.de/home/macrosymp

The service is free for everybody.

Contents of Macromol. Symp. 142

4th German-Korean Symposium
"Advanced Polymeric Materials – New Synthetic Routes and Applications"
Dresden, Germany, 1998

* The asterisk indicates the name of the author to whom inquiries should be addressed.

Author Index

Preface

Highlighted by the aspects of rapidly growing polymer industry and polymer research of Korea and well grown-up scientific and industrial collaboration between Korea and Germany, the need of a joint polymer symposium was first discussed in 1989 and resulted to the first bilateral symposium on development and trends in Polymer Materials Science organized in Seoul in 1991 by coordination of Prof. Dr. Man Jung Han and Prof. Dr. Gerhard Wegner. The follow-up symposium was held in 1993 at Max-Planck-Institute for Polymer Research in Mainz. The two very successful symposia have led to the conclusion that the further discussion of research programs, results, trends and be proceeded with young scientists with fresh ideas and recent research results.

The third symposium on New Synthetic Polymers and Complex Systems of Polymers was organized by Prof. Jin Chul Jung and Prof. Dr. Matthias Ballauff under mutual agreement between the Korean Science and Engineering Foundation (KOSEF) and Deutsche Forschungsgemeinschaft (DFG) in 1995 at the Pohang University of Science and Technology.

The 4th German-Korean Symposium was titled "Advanced Polymeric Materials – New Synthetic Routes and Applications" and was organized by Prof. Dr. Hans-Jürgen Adler at the Technische Universität Dresden.

In this symposium, 11 Korean and 12 German presentations gave an excellent overview about present research fields in both countries, from modification of well-known polymers to new polymer structures, like hyperbranched and dendrimeric polymers, from new surface modification to application in the microsystem technique and biocompatibility of polymers. Further fields such as optimization of optical properties, liquid crystalline polymers and rod-like polymers were also discussed.

This conference should be not the last and will be extended in successful collaboration.

The participants are very thankful for sponsorship by DFG, KOSEF, BASF AG and Bayer AG.

Hans-Jürgen Adler

Polyethyloxazoline Monolayers for Polymer Supported Biomembrane Models

Thomas Lehmann and Jürgen Rühe*

Max-Planck-Institute for Polymer Research, Ackermannweg 10, 55128 Mainz, Germany

Abstract

The synthesis of polymer monolayers, which are designed to act as "cushions" for lipid membranes attached to solid surfaces, is described. Monolayers of poly(ethyloxazolines) which could act as a polymeric support for biomimetic membranes have been attached to silicon oxide or gold surfaces following different strategies for the surface-attachment of the polymer chains. The polymers are either bound to the surface through a chemisorption process ("grafting-to") or "grown" on the surface of silicon oxide substrates through a cationic ring-opening polymerization started from the surface of the substrate, creating a surface-attached polymer monolayer *in situ* ("grafting from"). The characterization of the surface attached films by various surface analytical techniques and the swelling behavior of the surface-attached polymer films in moist air are described. The non-specific adsorption of fibrinogen as a typical example for a strongly adsorbing protein onto the surface attached monolayers is studied.

Introduction

Huge efforts have been directed in the last years towards the development of biosensor devices [1]. Such devices are designed to detect very small quantities of biologically or biomedically relevant molecules, which are present in an analyt such as blood for example, even in the presence of chemically or structurally very similar compounds [2-10] and in the presence of compounds, which attach unspecifically to the device surface ("surface fouling"). Accordingly large programs in organic chemistry have been directed towards the synthesis of very complex receptor molecules, which are supposed to react with the desired target molecule in the analyt with high specificity and reliability.

An alternative approach to the problem is to accept mother natures 4 billion years head start on the design and construction of highly specific receptors and use molecules of biological origin instead of synthetic ones. In this case the receptors, mostly proteins, have to be attached

* To whom correspondence should be addressed

 CCC 1022-1360/99/$ 17.50+.50/0

to the surface of a transducer, which transforms the chemical signal of the receptor into an electrical signal. This signal can then be processed further by conventional electronics. A promising way to provide a hospitable environment for such receptor proteins and an approach to orient them in an appropriate way at the surface of the device is to incorporate them into a membrane system, which tries to mimic the plasma membrane of a biological cell. An important task for the design of such a system is to carefully balance the stability and the fluidity of the membrane system. To allow for sufficient mechanical stability the attachment to a solid substrate seems advantageous, however a very rigid immobilization at the surface of a solid substrate will prevent movement in the membrane and prevent the receptor from working properly. To allow for such a compromise between fluidity and stability the concept of a polymer supported biomembrane has been developed.

The polymer layer, which supports the membrane, acts as a molecularly thin buffer between the membrane and the surface of the substrate [1]. It prevents contact between the incorporated proteins and the solid substrate, because such a contact could cause severe denaturing of the receptor molecules. Additionally the polymer "cushion" allows for the adjustment of the mechanical properties of the attached layers. On the one hand the tethering to the solid substrate anchors the membrane to the support and holds the whole assembly in place. It provides mechanical stability to the system and prevents the layer from collapsing of floating off. On the other hand having a soft "cushion" between the immobile substrate and the membrane decouples them and allows for sufficient mobility in the membrane system [2-10].
Important parameters for the properties of the "cushions" are accordingly the swellability of the polymer support in an aqueous environment and the interaction with other polymer molecules or low molecular weight compounds. It has been shown using other polymer monolayer systems, that the polymer conformation can have very profound effects onto the swelling behavior of the films or onto the wetting behavior of such systems [11]. Variations in the graft density can influence the degree of swelling of polymer monolayers, which gives the ratio of the dry vs. the swollen layer, by more than an order of magnitude. Additionally the surface-attachment of polymer molecules can influence the entropic situation in such a case that strongly, that the surface attached monolayers are not wetted even by chemically identical polymer ("autophobic" behavior) [12]. If such a monolayer is coated by an identical layer of non-attached polymer the physisorbed material will form more or less quickly droplets and dewet the underlying polymer coated substrate.

In this paper we focus only on the developement of different strategies for the build-up of the polymeric support layer. We describe the synthesis and characterization of monomolecular layers of poly(2-ethyl-2-oxazoline)s attached to planar, solid substrates. The polymers were chosen as they are hydrophilic, are known to be protein compatible and allow the reaction with lipids and groups, which immobilize the polymer at the surface of substrate. As substrate materials gold and silicon dioxide layers on silicon wafers are employed. We report on the layer thickness and graft density of the surface attached monolayers as well as the swelling behavior of the attached layers in moist air.

Results and Discussion

To allow for the attachment of the poly(ethyloxazoline) monolayers to the surfaces of gold substrates, polymer molecules with disulfide anchor groups had to be synthesized as shown in figure 1. Details of the synthesis and the characterization of the monolayers has been published elsewhere [13,14].

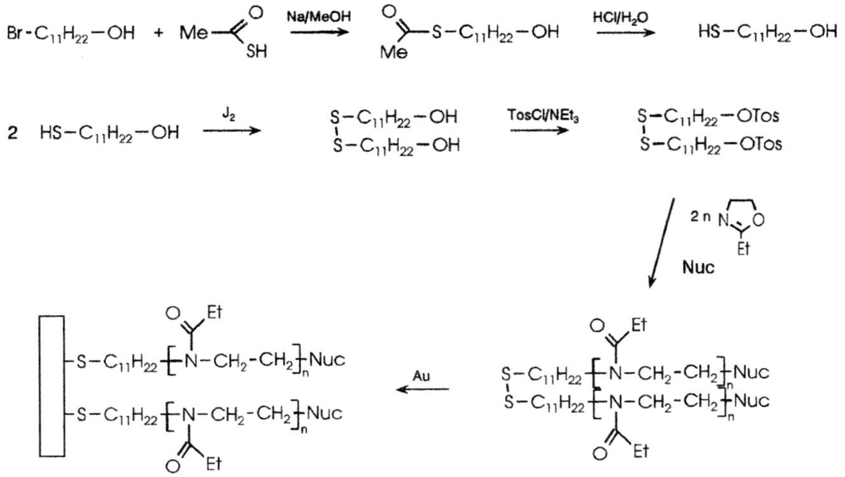

Fig. 1: Synthesis of poly(ethyloxazoline) monolayers attached to gold surfaces

To obtain the desired target polymers firstly tosylates of ω,ω'-dihydroxy alkyl disulfides were synthesized, which were used in a subsequent reaction step as initiators to start a cationic ring opening polymerization of 2-ethyl-2-oxazoline. The initiator was synthesized starting from 11-

bromo undecanol. The bromide was converted to a thiol via nucleophilic substitution with thioacetic acid and hydrolysis of the resulting thioester. The latter was oxidized with iodine to the corresponding disulfide-ω,ω'-diol and tosylated to give bis-11,11'-tosylato-undecyldisulfide, the desired initiator.

This was then used as a bifunctional initiator to start a cationic ring-opening polymerization of 2-alkyl-Δ2-oxazolines in solution. It has been shown, that, if the alkyl spacer connecting the disulfide moiety and the initiating site is long enough to ensure that the structural integrity of the disulfide is conserved during the polymerization process. A series of several poly(ethyl-oxazolines) with varying degrees of polymerization (P_n=5-150) and different endgroups (piperidine, dioctadecylamine and hydroxy) were synthesized (fig. 1). The endgroups were attached after consumption of all monomer by capping the living oxazolinium group with a nucleophile.

The polymers were chemisorbed from ethanol solution onto gold surfaces. The layer thicknesses of the monolayers after Soxhlet extraction were determined by surface plasmon resonance spectroscopy (SPR). With increasing length of the polymer molecules the layer thicknesses increases, but rapidly levels off at about 3.5 nm (Fig. 2).

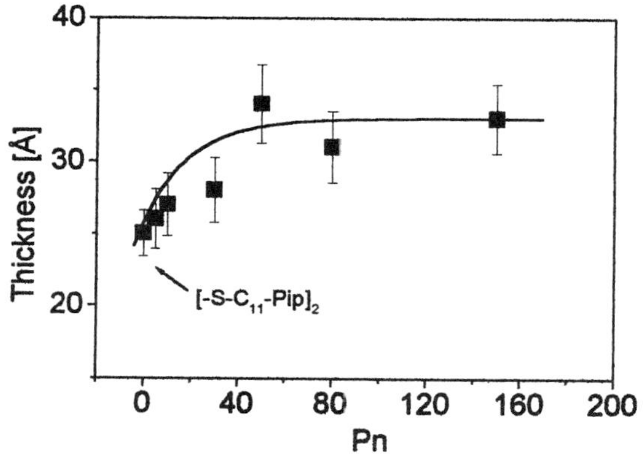

Fig.2: Thickness of chemisorbed poly(ethyloxaline) monolayers at gold surfaces as a function of the degree of polymerization P_n

As the layer thickness remains constant upon increasing of the chain length of the attached polymer molecules, the graft density of the chains must consequently be reduced with increasing molecular weight of the surface-attached chains. The coupling between graft density and molecular weight is intrinsic to all chemisorption processes of polymers to solid surfaces. The reason for such a behavior is as described in the following. Polymer molecules arriving at a later stage of the chemisorption process have to diffuse against the concentration gradient, build up by the chains attached in the early stages of the process, in order to successfully bind to the surface. Once a certain critical segment density within the surface layer is reached, the surface-attachment of the polymer molecules slows down due to this limitation of diffusion and further attachment of chains takes place only on a very slow (i.e. logarithmic) time scale [15] (Fig.3). This kinetic limitation reduces the film thickness which can be obtained under practical experimental conditions to values which are typically between 3 and 5 nm. If now longer polymer chains are used for the chemisorption, the critical segment density at which the kinetics starts to control film growth, is simply reached at a lower graft density. In this case the layer thickness becomes independent of the length of polymer molecules.

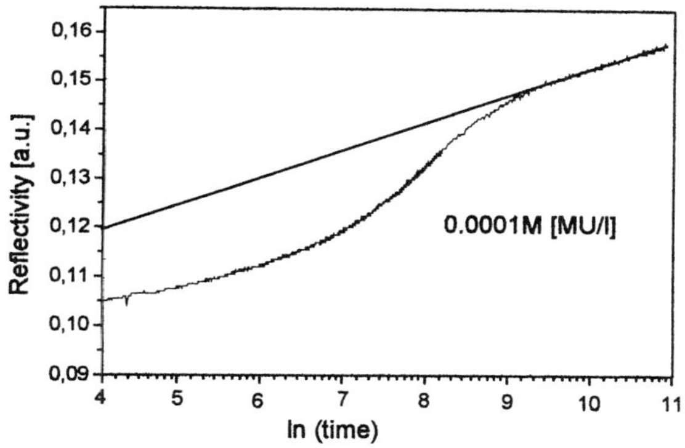

Fig. 3: Kinetics of the surface attachment of a disulfide modified poly(ethyloxazoline) at a gold surface as measured by SPR experiments

The number of molecules attached to the gold surface per surface area can be directly determined if the polymer chains are electrochemically cleaved off from the surface of the substrate and the number of charges flown during this process is determined. In Figure 4 cyclic voltammetry (CV) experiments are shown, in which polymers of different degrees of polymerization were attached to gold electrodes. From the integral of the reduction peak in the CV experiment the number of charges consumed and thus the number of cleaved Au-S bonds can be directly inferred. It is evident that with increasing chain length of the polymer the area under the peak and thus the number of surface-attached polymer molecules is strongly reduced.

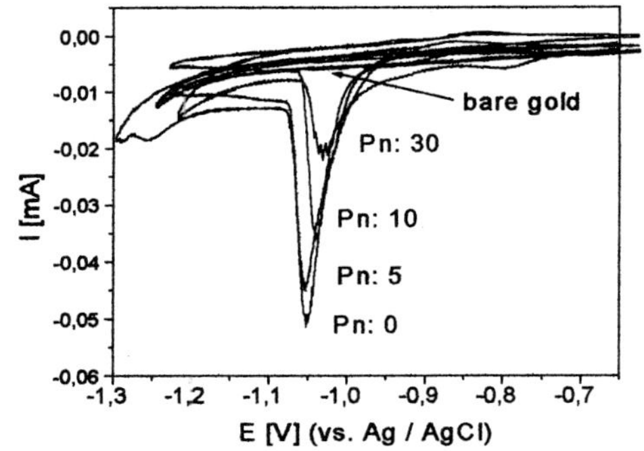

Fig. 4: Cyclovoltammogram of poly(ethyloxazoline) monolayers attached to gold electrode surfaces

One important question for the reproducibility of the synthesis of such systems is how the film thickness is influenced by the concentration of the polymer molecules in solution.

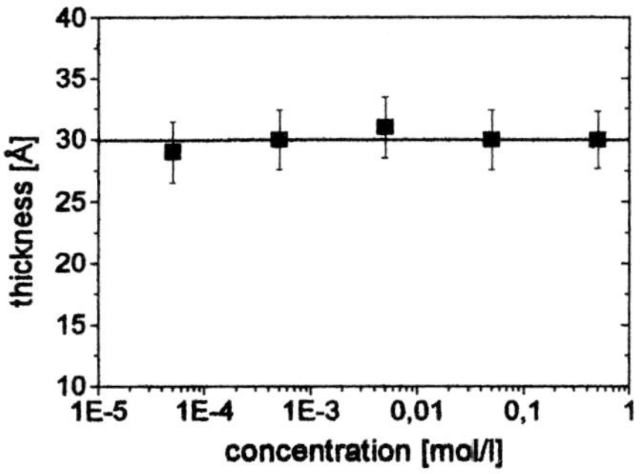

Fig. 5 Thicknesses of monolayers of disulfide modified poly(ethyloxazoline) self assembled from ethanol solution onto a gold substrate at different polymer concentrations

In Figure 5 the thicknesses of monolayers of one polymer are shown as a function of the concentration. It is clearly visible, that the layer thickness is not influenced by the concentration of polymer molecules present during the chemisorption reaction. This contrast strongly to the behavior of the surface-attachment of partially hydrolyzed poly(ethyloxazolines) attached to surfaces modified with appropriate "anchor" molecules in a step-wise chemisorption process resulting in multiple point anchoring of the polymers to the surface (Fig. 6).

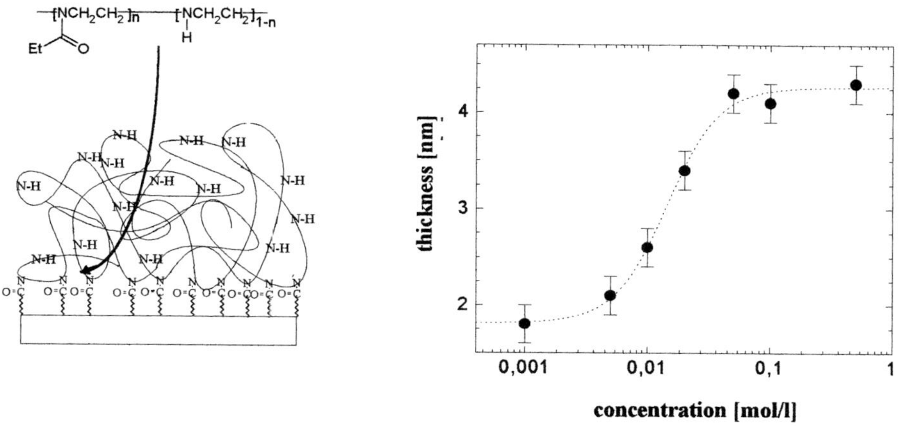

Fig. 6: Thickness of polyoxazoline monolayers obtained by a sequential approach in which poly(ethyloxazoline-stat-ethyleneimine) polymers are attached to N-hydroxy-succinimide self-assembled monolayers

The concentration dependence in the latter case can be explained if at first a situation is envisaged, were one reactive segment of the polymer molecule is attached to a corresponding site at the substrate (Fig. 6a). Other reactive groups, which are part of the same polymer molecule, can now also get into contact with the surface and become attached. As the surface attachment reaction is irreversible, the polymer is "frozen" into a non-equilibrium, pancake like structure. The polymer "rolls out" on the surface and many different segments of the same chain become attached to the surface. As all reactive sites at the surface of the substrate become consumed the attachment of further molecules from solution is prevented.

If in contrast to this the chemisorption process is carried out in highly concentrated solutions segments of other polymer molecules efficiently compete for the reactive sites on the surface. The number of attached molecules per surface area raises and consequently the film thickness increases. This increase, however, saturates at a certain value as at that point the thickness of the layer is no longer limited by the number of available reactive sites at the surface, but diffusion of the polymer molecules through the attached chains becomes the limiting factor as already discussed above. Thus an S-shaped curve is obtained when the layer thickness is plotted as a function of solution concentration (Fig. 6b).

It is important to note, that in both cases described above (chemisorption in one step or stepwise surface-attachment) the polymer is essentially the same, only the conformation of the chains at the surface is different.

"Growth" of poly(ethyloxzaolin) brushes

Independent of the exact nature of the chemisorption process the thicknesses all polymer layers obtained by such a "grafting-to" process are intrinsically limited as already discussed above. Due to kinetic reasons the layer thickness of such monolayers is restricted to a few nanometers for all practical experimental time scales. If higher film thicknesses are required the polymer molecules have to be "grown" at the surface of the substrate *in situ* [13]. To achieve this, a self-assembled monolayer of a tosylate initiator is formed at the surface of a substrate and a cationic ring-opening polymerization is started from this surface-attached monolayer (Fig. 7). In order to obtain a well defined ratio between monomer and initiator concentration "free" initiator is also added, which starts the generation of polymer molecules, which are not attached to the substrate surface.

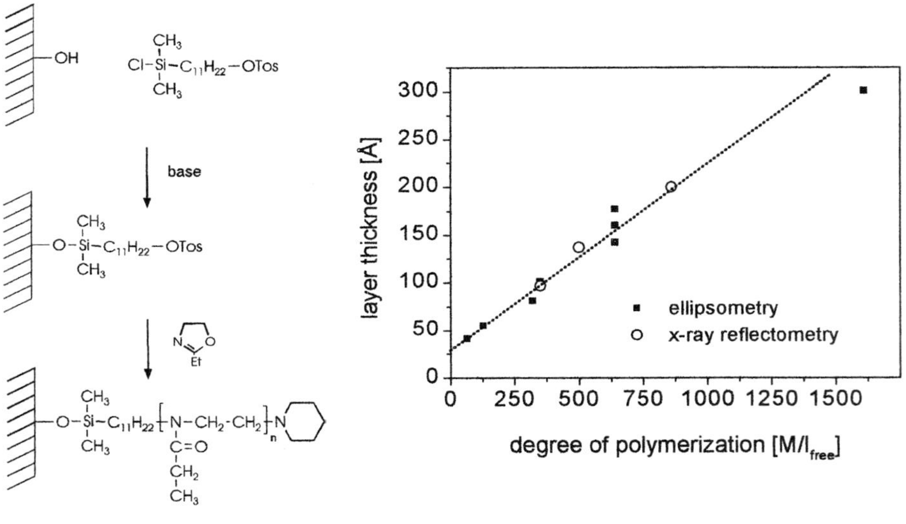

Fig. 7: a) Synthesis of poly(ethyloxazoline) monolayers by a "grafting-from" procedure; b) film thicknesses of the obtained layers

Following this "grafting-from" approach polymer monolayers of 20 - 30 nanometers can be synthesized easily. Thus the layers are about one order of magnitude higher than those, which can be obtained using a simple chemisorption process. The layer thicknesses of the monolayers can be controlled by adjusting the monomer/initiator ratio (Fig. 7b) as expected for such a polymerization reaction.

Swelling of the Surface-attached Monolayers in Humid Air

One very important property of the synthesized polymer monolayers is how strongly the layers are swollen in an aqueous environment as analysis of biologically active molecules is usually carried out in water. As a first experiment into this direction the swelling of the layers in humid air was studied. The polymers were brought into contact with air, whose relative humidity was controlled by placing small vessels with concentrated salt solutions into the closed measurement cell. The equilibrium water vapor pressure at room temperature was controlled by using different salt solutions. The thickness of the layers was measured by SPR spectroscopy in situ (Fig. 8).

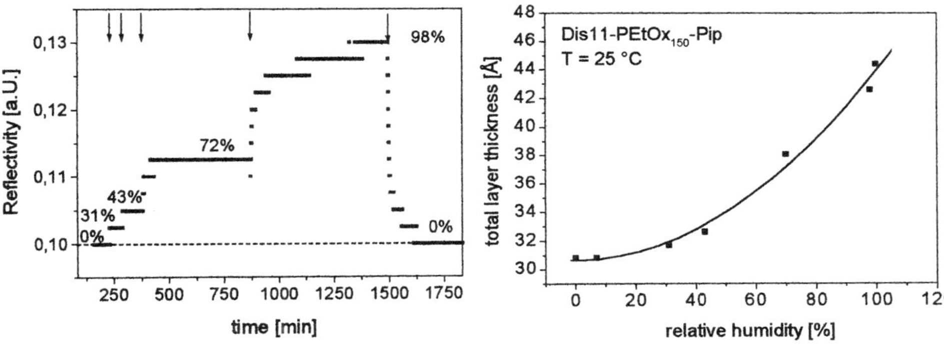

Fig. 8: Swelling of poly(ethyloxazoline) monolayers covalently attached to gold surfaces as a function of the relative humidity at room temperature; layer synthesis according to Fig. 1; a) SPR measurements, arrows indicate change of humidity; rel. humidity values are given in the graph b) thickness values obtained from SPR measurements

It can be seen, that the layer thickness increases up to 50% upon exposure to humid air (Fig. 8b).If poly(ethyloxazoline) brushes are employed even larger absolute increases of the film thickness due to water uptake in moist air can be realized (Fig.8). Experiments on the measurement of the film thickness of the layers, which have been brought into contact with bulk water are currently being carried out.

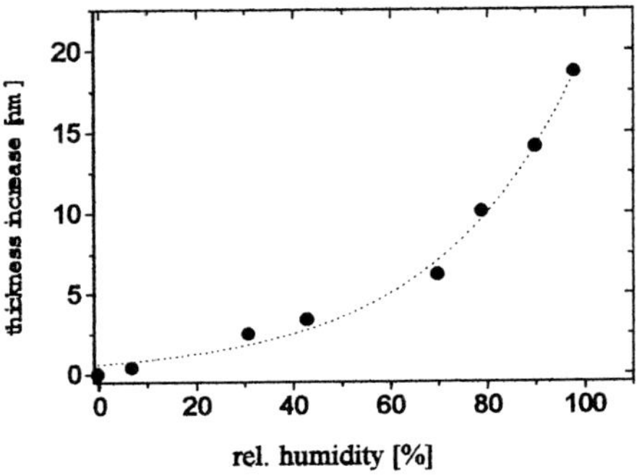

Fig. 9: Swelling of poly(ethyloxazoline) brushed in air having different relative humidities at room temperature; layer thickness at 0 % rel. hum. = 20 nm

Protein Adsorption Studies

As pointed out already above the resistance of the polymer layers against non-specific adsorption of proteins from the analyt is a very important requirement for the properties of a membrane support. If strong adsorption of proteins takes place the device will be covered after a short time with a thick layer of proteins, which limit access to the receptors for the molecules to be detected. To study the non-specific adsorption of proteins at the monolayers fibrinogen was chosen as a model compound, as it is known, that fibrinogen is a particular "sticky" protein, which adsorbs strongly at the surfaces of many different substrates. All experiments were carried out in a 0.1 % solution of the protein in phosphate buffer. The thickness of the protein layer was measured by SPR in situ. At first monolayers of the polymers having different degrees of polymerization were prepared on gold surfaces and brought into contact with the buffer solution (0.02 M phosphate buffer, pH = 7.5). After recording of an SPR spectrum the solution was exchanged against another part of the same solution, however, where fibrinogen has been added. The increase of the film thickness due to adsorption of the protein was followed (Fig. 10). When the film thickness did not increase any longer the sample was rinsed with pure buffer and again an SPR spectrum was recorded. From the angular shift before and after protein exposure the amount of adsorbed protein was calculated.

With increasing length of the polymer molecule the influence of the relatively hydrophobic initiator fragment, which remains in the polymer after completion of the polymerization process, decreases and the amount of adsorbed fibrinogen decreases. Indeed, when a polymer with a degree of polymerization of P_n = 150 was employed only 0 - 0.3 nm of fibrinogen could be adsorbed and the polymer coated surfaces were almost completely resistant against non-specific adsorption of proteins.

a)

b)

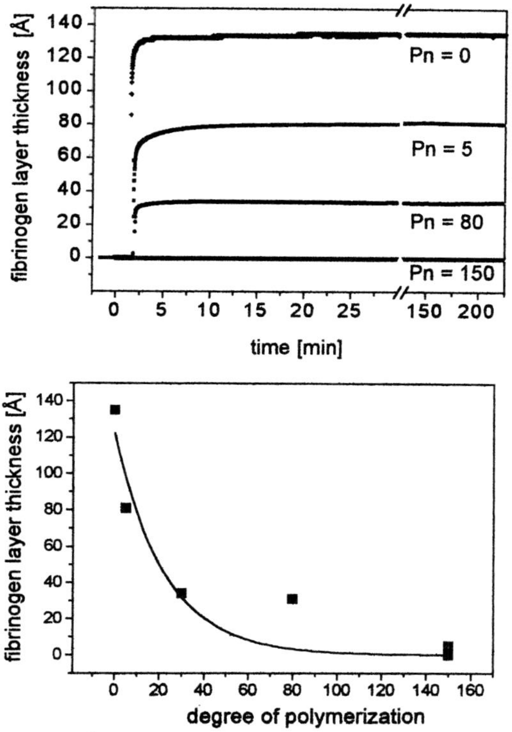

Fig. 10: adsorption of fibrinogen onto poly(ethyloxazoline) monolayers with varying degrees of polymerization; the layers are covalently linked to gold substrats; a) layer thickness as a function of adsorption time as measured by SPR; b) layer thickness as a function of degree of polymerization.

Conclusions

Poly(ethyloxazoline) monolayers have been prepared at the surface of silicon oxide and gold surfaces. While in the latter case the surface-attachment reaction is achieved through disulfide groups contained in the polymer molecules, in the former case the covalent bonds between polymer and substrate are generated through silane chemistry. The structure of the polymer chains, and accordingly the physical properties of the films, depend very strongly on the conditions of monolayer deposition. Under some conditions (multiple anchor points) very flat conformations of the polymer molecules are obtained. If other monolayer preparation protocols are employed (i.e. "growth" of polymer molecules at the surface) the polymer chains are strongly stretched away from the surface. Accordingly film thickness and graft density of the polymer chains can be fine-tuned in a wide range. This allows to study directly the influence of the conformation of the chain "tethered" to the surface onto the macroscopic behavior of the monolayers.

Acknowledgments

T.L. wishes to thank the foundation of the Chemical Industry ("Stiftung Stipendienfonds im Verband der Chemischen Industrie") for a scholarship. Dr. A. Badia is thanked for carrying out the cyclic voltammetry measurements.

References

[1] E. Sackmann, *Science,* **271**, 43 (1996)
[2] B.A. Cornell, V:L.B. Braach-Maksvytis, L.G. King, P.D.J. Osman, B. Raguse, L. Wieczorek and R.J. Pace, *Nature,* **387**, 580 (1997)
[3] H. Ringsdorf, B. Schlarp, J. Venzmer, *Angew. Chem.* **100**, 117 (1988)
[4] A. Arya, U.J. Krull, M. Thompson, H.E. Wong, *Anal. Chim. Acta,* **173**, 331 (1985)
[5] C. Dietrich, R. Tampe, *Biochim. Biophys. Acta,* **1238,** 183, (1995)
[6] M. Uto, M. Araki, T. Taniguchi, S. Hoshi, S. Inoue, *Analyt. Sci.* **10**, 943 (1994
[7] R. Naumann, A. Jonczyk, R. Kopp, J. Esch, H. Ringsdorf, W. Knoll, P. Gräbner, *Angew. Chem.* **107**, 2168 (1995)
[8] H.M. McConnell, T.H. Watts, R.M. Weis, A.A. Brian, *Biochim. Biophys. Acta,* **864,** 95 (1986)
[9] H. Lang, C. Duschl and H. Vogel, *Langmuir,* **10**, 197 (1994)
[10] C. Heibel, S. Maus, W. Knoll and J. Rühe, *ACS Symp. Ser.* **695**, 104 (1998)
[11] a) Halperin A., Tirell M., Lodge T.P. in *Advances in Polymer Science,* 100, Springer Verlag Berlin, Heidelberg (1992) b) S.T. Milner, Science **251**, 905 (1991)
[12] a) G. Reiter, P. Auroy, L. Auvray, *Macromolecules,* **29**, 2150 (1996); b) G. Henn, D. Bucknall, M. Stamm, P. Vanhoorne, R. Jerome, *Macromolecules,* **29**, 4305 (1996); c)E. Ponomarenko and J. Rühe, submitted for publication
[13] a) O. Prucker and J. Rühe, *Langmuir* **14**, 6893 (1998);b) J. Rühe, *Macromol. Phys. Chem., Macromol. Symp.* **126**, 215 (1997)
[14] T. Lehmann, Ph.D. thesis, University of Mainz, 1997
[15] T. Lehmann and J. Rühe, submitted for publication
[16] a) R. Zajac and A. Chakrabarti, *Phys. Rev. E,* **52**, 6536 (1995); b) Kopf,A., Baschnagel, J., Wittmer, J. and Binder, K., *Macromolecules,* **29**, 1433 (1996)
[17] R. Jordan and A. Ulman, *JACS,* **120**, 243 (1998)

Effect of Molecular Orientation on the Biodegradability of Aliphatic Polyester

Eui Sang Yoo, Seung Soon Im*

Department of Textile and Polymer Engineering,
Graduate School of Advanced Materials and Chemical Engineering,
Hanyang University, Hangdang-dong, Sungdong-gu, Seoul 133-791, KOREA

SUMMARY: The effect of morphological microstructure on the biodegradability of aliphatic polyester, poly(ε -caprolactone) (PCL) was studied in terms of crystallite size, crystallinity and amorphous and crystalline orientation factors. Microstructural changes during hydrolysis/biodegradation of the drawn PCL films were investigated by the conventional small and wide angle X-ray scattering methods. The lower was the draw ratio, the higher the hydrolytic degradability or biodegradability. With the increase of the hydrolysis time, the long period, at earlier stage, decreased; then slightly recovered and the crystallinity increased while the lamellar thickness remained unchanged. The amorphous orientation factors start to decrease at earlier stage and gradually go down to zero before the end of hydrolysis. In the case of crystalline orientation factor, although the values decrease with increasing hydrolysis time, they do not reach zero point. After the biodegradation for 60 days, crystallinity, crystal lateral size and lamellar thickness in all drawn PCL films decreased, and then it was confirmed that even crystalline regions were degraded for long term biodegradation test.

Introduction

Aliphatic polyester has been known as the environmental biodegradable polymer and there were so many reports on degradation mechanisms of aliphatic polyesters and their blends or copolymers with other common polymer. It is well known that hydrolytic attack could be initiated in the amorphous regions of the polymer[1-5]. The amorphous regions of polymer are more open and less dense than the crystalline regions and are more susceptible to be attacked by reacting species or solvent than the crystalline regions. By the way, it has been reported that the biodegradability was improved with not only the increase of amorphous contents but also the lower melting temperature due to low perfectness of crystals[6-8]. The biodegradability of crystalline polymers usually depends on the crystalline, amorphous microstructure such as size and perfectness of crystallites, crystallinity and degree of orientation[9-11]. There were, however, few reports on the relationship between microstructure and biodegradability.

 CCC 1022-1360/99/$ 17.50+.50/0

The aim of this work is to identify the effect of morphological microstructure in the amorphous and crystalline region on the biodegradability of aliphatic polyester, poly(ε -caprolactone) (PCL). The molecular orientations in crystalline and amorphous regions of drawn PCL (poly ε -caprolactone) were measured by the X-ray azumithal scanning method and the microstructural changes during degradation of the drawn films were investigated by the conventional small and wide angle X-ray scattering methods.

Experimentals

The PCL films were prepared by quenching in the cold water after being melt-pressed. The films, then, were drawn at the temperature of -55℃ which is a little higher than the glass transition temperature of PCL. The quenched film thickness was controlled so that the drawn films had the almost same thickness, because the surface area of samples is a very important factor of the degradation test. The different morphological parameters depending on two degradation test methods - the biodegradation method in activated sludge and the hydrolytic degradation method – were investigated. By using the wide angle X-ray azimuthal scan, the molecular orientation of amorphous and crystalline regions and the crystallinity of drawn samples were measured. A small angle X-ray diffractometer was used for investigating the variation of long period during the degradation.

Results and discussion

Tab. 1 denotes the fine structures of drawn PCL films measured by wide- and samll angle X-ray diffractometer. In the meridional SAXS intensity profiles (lamellar intensity) with draw ratio, the maximum of peak intensity moves to lower angle and the lamellar intensity increases with increasing draw ratio. It is well known that the increase of SAXS intensity is usually attributed to an increase in the crystallinity and crystalline order, and a decrease in interlamellar density[12,13]. In Tab. 1, the long period increases with draw ratio. From the value of long period and crystallinity, the lamellar thickness was calculated[14,15]. With increasing the draw ratio, the size of crystallite along the draw direction, lamellar thickness, appears to increase and that perpendicular to the draw direction to decrease. The crystallinity slightly increases as the draw ratio increases. WAXS profile obviously shows two sharp peaks, 110 plane and 200 plane reflections. The d-spacing of 110 and 200 planes was not changed by drawing PCL films. That is, drawing does not affect the crystal structure.

The variations of crystal orientation factor, f_c, and amorphous orientation factor, f_{am}, with drawing ratio were presented in Tab. 1. Orientation factor f_c drastically increases at the lower elongation and then the increment rate is reduced with further elongation. This is the typical trend of the oriented polymer.

Tab. 1. Fine structure parameters of drawn PCL films measured by wide- and small-angle X-ray diffractometer

Draw Ratio	Crystallinity (%)	f_c	f_{am}	Crystallite Size (Å)			Crystal Volume (Å^3)
				L (110)	L (200)	Lamellar Thickness	
1 (undrawn)	34.3	0	0	73.6	68.7	57.6	349,500
2	35.1	0.48	0.28	68.6	65.5	61.4	331,000
3	35.3	0.69	0.56	67.7	63.8	64.2	332,700
4	37.8	0.81	0.71	65.8	61.8	70.6	344,500
5	40.6	0.84	0.79	64.9	60.5	76.7	361,400

Fig. 1 shows the variation of degradability of PCL films with different draw ratio as a function of hydrolysis time. After 20 days hydrolysis, the films could not keep their shape so that the measurement of weight was impossible. The lower the draw ratio, the higher the degradability and the faster the starting of notable degradation. The molecular density in the amorphous region becomes high as the orientation increases. The higher molecular density might interrupt the penetration of active species between the molecular chains, resulting in the delayed starting point of the degradation and the less degradability when the draw ratio is high.

The characteristics of SAXS intensity profiles for the PCL films, such as the position of maximum intensity, curve shape and maximum intensity, depended on the hydrolysis time. The position of maximum intensity moved to higher angle with hydrolysis time, which means the decrease of the long period. The lamellar intensity increased with increasing hydrolysis time due to the increase of electron density difference between crystalline and amorphous regions by hydrolysis. The long periods were calculated and plotted versus hydrolysis time in Fig. 2. In the case of the films with 3 and 4 draw ratio, the values of long period, at earlier stage, decrease, and then slightly recover. The same phenomenon was reported by R. E. Cameron[16] for the poly(glycolic acid), even though undrawn, molded

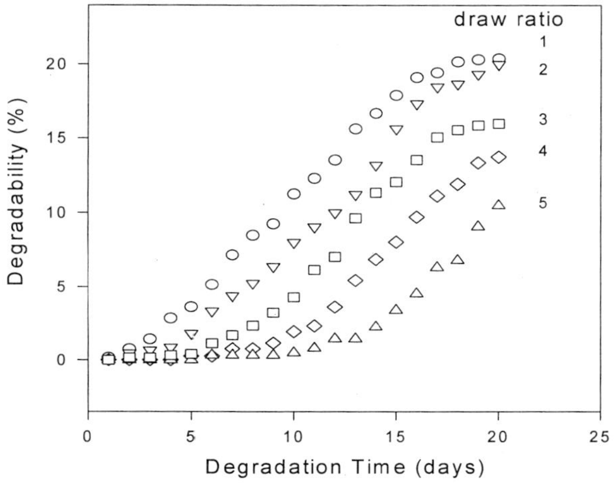

Fig.1. The variation of degradability of the PCL films with various draw ratio. The degradability by hydrolysis was calculated from the weight loss.

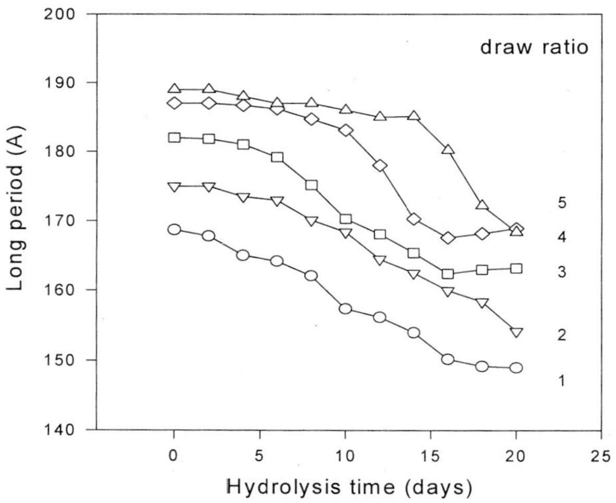

Fig.2. The variation of long period for the PCL films with hydrolysis time. The values are obtained by applying the Bragg equation to the peak positions of the Lorentz corrected SAXS intensity profiles.

sample was used. Our undrawn or little drawn samples showed no recovery of long period with the hydrolysis time. In the process of hydrolytic degradation, the molecular chains in amorphous regions were selectively attacked and degraded. Accordingly, the long period was reduced because the tie molecules in amorphous regions were removed. However, at the final stage of hydrolysis, the long period was slightly increased due to the volume increase of degraded amorphous chains and voids. The reason this phenomenon could not be observed in the undrawn or low-drawn samples might be that they have comparatively low density in amorphous region. .

Fig. 3 shows the crystallinity variation with different draw ratio as hydrolysis time increases. It is sure that the crystallinity increases due to the removal of molecular chains in amorphous region. After 20 days hydrolysis, the crystallinity reached reasonable values, meaning that these crystallinities were a little lower than those calculated from the weight loss after 20 days hydrolysis in the assumption that the whole weight loss corresponds to the amorphous regions. On the proceeding of the hydrolysis, the variation of lamellar thickness was traced by using the crystallinity of each sample measured from WAXS intensity profiles and long period from SAXS intensity profiles (Fig. 4). Lamellar thickness remained unchanged for all the samples during hydrolysis.

Even though it was possible that the constraints between the crystallites could be reduced as the tie molecules in amorphous region were removed, the effect of degradation on the unit cell dimensions was almost negligible in this work. That is, the crystal density was unchanged by the degradation. We could get some information about average lateral crystal size from the width of the WAXS peaks 110 and 200 using Debye Scherrer equation. The slight decrease in the width of 110 and 200 is observed with hydrolysis time. It means that the average lateral crystal size increased because so smaller crystallites were more readily removed due to their large surface area.

In Fig. 5, the changes of amorphous orientation factors were represented with the hydrolysis time. The amorphous orientation factors start to decrease at earlier stage and continually go down to zero before the end of hydrolysis, meaning that the orientation of molecular chains in amorphous region randomized from the earlier stage of hydrolysis. In the case of crystalline orientation factor in Fig. 6, although the values decrease with increasing hydrolysis time, they do not reach zero point. It means that the crystallites keep their orientation to some extent after hydrolytic degradation.

The biodegradability was evaluated by the amount of CO_2 creation from PCL in the activated sludge. The biodegradation behavior is similar to the results of hydrolytic degradation. The lower was the draw ratio, the higher the degradability. Tab. 2 denotes the change of fine structure for drawn PCL films measured by WAXS and SAXS after

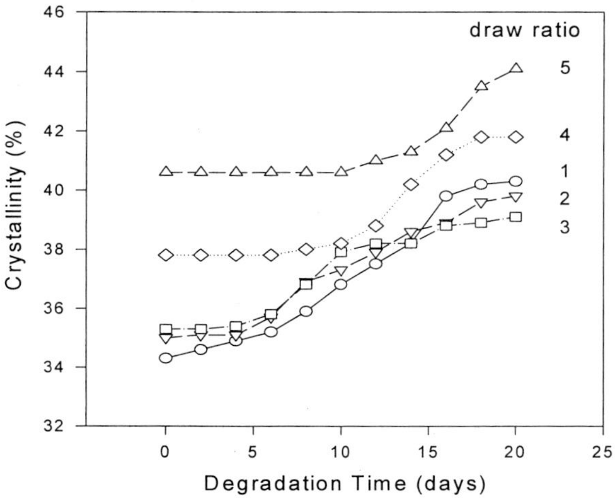

Fig.3. The crystallinity of the PCL films calculated from the WAXS intensity profiles for samples with different draw ratio.

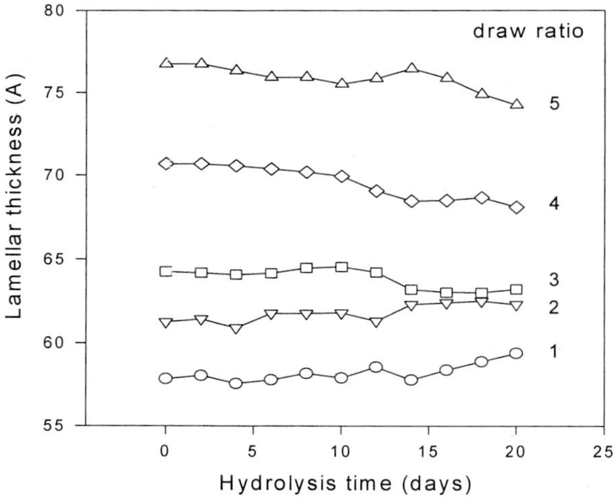

Fig.4. The variation of lamellar thickness for the PCL film with hydrolysis time. The thickness were calculated from the values of long period and crystallinity

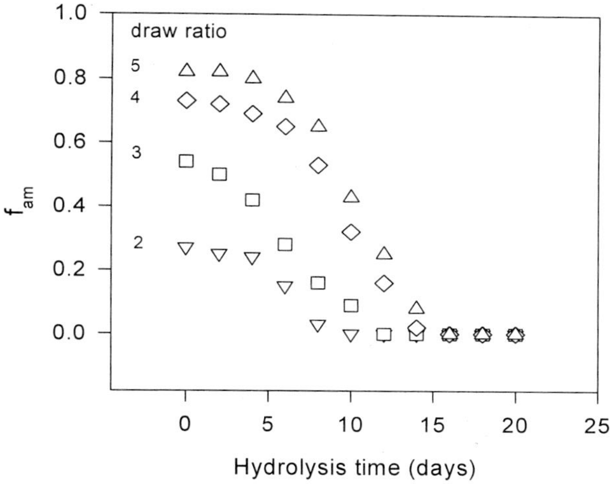

Fig.5. The variation of amorphous orientation function for PCL films depending on the hydrolysis time.

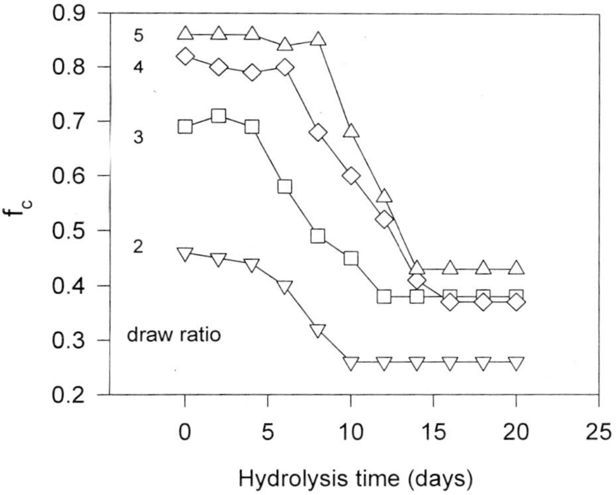

Fig.6. The variation of crystalline orientation function for PCL films depending on the hydrolysis time.

biodegradation in activated sludge. In the course of hydrolytic degradation, we suggested that the amorphous regions were mainly degraded including some very small and imperfect crystallites. So, the average lateral crystallite size increased with hydrolysis time. After the biodegradation for 60 days, however, crystallinity, crystal lateral size and lamellar thickness in all drawn PCL films decreased. Therefore, we confirmed that even crystalline regions were degraded for long term biodegradation test.

Tab. 2. Fine structures of drawn PCL films before and after biodegradation in activated sludge measured by wide- and small-angle X-ray diffractometer

Draw Ratio	Degradation Test	Quenched film				
		Crystallinity (%)	Crystallite Size (Å)			Crystal Volume ($Å^3$)
			L (110)	L (200)	Lamellar Thickness	
1	Before	34.3	73.6	68.7	57.6	349,500
	After	31.8	69.2	63.1	53.1	277,800
2	Before	35.1	68.6	65.5	61.4	331,000
	After	33.6	62.1	61.3	55.4	253,000
3	Before	35.3	67.7	63.8	64.2	332,700
	After	32.1	62.2	58.9	60.1	262,900
4	Before	37.8	65.8	61.8	70.6	344,500
	After	34.5	58.4	54.3	64.3	244,700
5	Before	40.6	64.9	60.5	76.7	361,400
	After	33.4	57.1	55.4	69.6	264,200

Conclusion

The effect of morphological microstructure on the biodegradability of aliphatic polyester, poly(ε -caprolactone) (PCL) was studied in terms of crystallite size, crystallinity and amorphous and crystalline orientation factors. The lower was the draw ratio, the higher the degradability. The values of long period, at earlier stage, decrease, and then slightly recover. The long period was reduced because the tie molecules in amorphous regions were removed. However, at the final stage of hydrolysis, the long period was slightly increased due to the volume increase of degraded amorphous chains and voids. The crystallinity increases due to the removal of molecular chains in amorphous region and lamellar thickness remained unchanged for all the samples during hydrolysis.

The amorphous orientation factors start to decrease at earlier stage and gradually go down to zero at the middle term of hydrolysis, meaning that the orientation of molecular chains in amorphous region randomized from the earlier stage of hydrolysis. In the case of crystalline orientation factor, although the values decrease with increasing hydrolysis time, they do not reaches to zero point. It means that the crystallites keep their orientation to some extent after hydrolytic degradation.

The behavior of the biodegradation in activated sludge is similar to the results of hydrolytic degradation. The lower was the draw ratio, the higher the degradability. After the biodegradation for 60 days, however, crystallinity, crystal lateral size and lamellar thickness in all drawn PCL films decreased. Therefore, we confirmed that even crystalline region was affected in the long termed biodegradation test.

Acknowledgements

This work was carried out under a research grant from the Korea Science and Engineering Foundation (# 95-0300-07-03-3).

References

[1] G. Swift, *Acc. Chem. Res.*, **26**, 105 (1993)
[2] R. Lenz, . *in Advances in Polymer Series*, **107**, (esd N. A. Peppas and R. S. Langer), Springer-Vierlanger, p.1 (1993)
[3] A. C. Albertsson, S. Karlsson, *in Comprehensive Polymer Science*, First Supplement (eds G. A. Allen, S. L. Agarwal and S. Russo), Pergamon Press, London, , p.285 (1993)
[4] E. S. Yoo, Y. Yoo, S. S. Im, *Macromol. Symp.* **118**, 739 (1997)
[5] E. S. Yoo, S. S. Im, *J. Environ. Polym. Deg.* 6, No. 4 (1998)
[6] E. W. Fischer, H. J. Sterzel, G. Wegner, *Kolloid-z. u. Z. Polymere* **251**, 980 (1973)
[7] B. K. Carter, G. L. Wilkes, S. W. Shakaby, A. S. Hoffman, B. D. Ratner, T. A. Horbett, "*Polymers as Biomaterials*", Plenum Press, New York, P.67
[8] R. J. Fredericks, A. J. Melveger, L. J. Dolegiewtz, *J. Polym. Sci.; Polym. Phys. Ed.* **22**, 57 (1984)
[9] S. J. Holland, B. J. Tighe, P. L. Gould, *J. Control. Rel.* **4**, 155 (1986)
[10] T. St Pierre, E. Chiellini, *Bioact., J. Comp. Polym.* **2**, 4 (1987)
[11] M. Vert, *Macromol. Chem., Macromol. Symp.* **6**, 109 (1986)
[12] R. Bonart, R. Hosemann, F. Motzkus, H. Ruch, *Norelco Reptr.* **7**, 81 (1960)
[13] R. J. Samuels, "*Structured Polymer Properties*", John wiley & Sons, New York,
[14] J. W. Leeslag, S. Gogolewski, A. J. Pennings, *J. Biomed. Mater. Res.* **29**, 2829 (1984)
[15] J. H. Hermans, P. H. Hermans, D. Bermaas, A. Weidinger, *Rec. Trv. Chim.* **65**, 427 (1946)
[16] E. King, R. E. Cameron, *J. Polym. Sci.* **27**, 990 (1997)

The Effect of Spacer Length on the Polymerization of Diacetylenes in SAMs on Gold Surfaces

Henning Menzel[1], Mark D. Mowery[2], Mei Cai[2], Christine E. Evans[2]

[1] Universität Hannover, Institut für Makromolekulare Chemie, Am Kleinen Felde 30, 30167 Hannover, Germany

[2] University of Michigan, Chemistry Department, 930 N. University Ave., Ann Arbor, MI 48109-1055, USA

SUMMARY: The influence of the molecular architecture of diacetylene disulfides on the structure and polymerizability of self assembled monolayers on gold was investigated. The position of the diacetylene group was varied systematically by adjusting the length of the tail and the spacer. For the strutures studied, the longer the tail, the higher the akyl chain crystallinity as sensed by IR spectroscopy, and the higher the polymerization efficiency as probed by resonance Raman spectroscopy. The influence of the spacer on the polymerizability is more complicated. It is conjectured that the spacer reduces the lattice strain due to the misfit of the arrangement of the anchor groups and the arrangement of the diacetylene units and the polydiacetylene backbone, respectively. Simple modeling reveals that the number of the methylene units (odd or even) and the overall length both influence the arrangement. The highest polymerization efficiency is found for the longest, odd numbered spacer.

Introduction

Poly(diacetylene)s (PDA) exhibit an electronic transition due to the extented π-system. Characterized extensively as crystals and LB films,[1-14)] polydiacetylenes are known to exist in several phases which appear blue, purple, or red in color. Although the exact nature of this chromism remains elusive, it is generally attributed to variations in the effective conjugation length, which is determined by the length of the polymer chain, the chain conformation and the local environment of the polymer backbone.[11)] Because the excitonic absorption is directly related to this effective conjugation length, the blue phase is characterized by the longest conjugation length with the purple and the red forms indicative of successively diminished conjugation length.[3-7,15,16] Furthermore, both reversible and irrreversible phase transitions, particularly between the blue and red phases, have been observed. PDAs in LB films[15,17,18)] or in liposomes[19,20)] have been suggested for application in colorimetric sensors. In these sensor designs, the conjugation length of the polymer is conjectured to be diminished upon interaction of the target molecule with the diacetylene bilayer, leading to a color change response. For such sensor applications it is therefore of great importance to obtain polymer with a long effective conjugation length, i.e. blue phase polymer. However, the constraints for the topochemical polymerization of diacetylenes are considerable[12)] and poly(diacetylene)s with a long conjugation length (blue phase polymer) can be obtained only if the monomeric assembly is in a highly ordered state which meets these constraints (see Figure 1).

 CCC 1022-1360/99/$ 17.50+.50/0

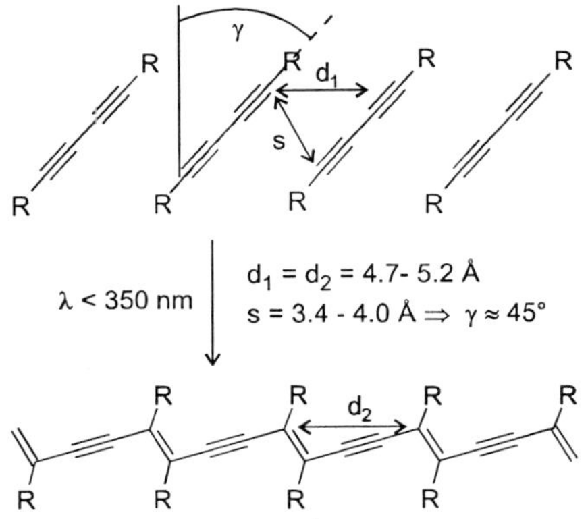

Figure 1: Steric requirements for the polymerization of diacetylenes.

For LB films incorporating diacetylene moieties, the interdependence between the molecular architecture and the structure and polymerizability is well studied.[3,5-9,11] Recent studies have shown that poly(diacetylene) assemblies also can be prepared as self-assembled monolayers (SAMs).[21-25] These SAMs are of particular interest because they are much more stable than LB films due to the covalent binding to the substrate.[26] Furthermore, the preparation of SAMs is not limited to planar substrates.[27] However, in self-assembled monolayers, the interaction between the headgroup and the surface is much stronger than in LB films due to the covalent linkages at the surface. As a result interdependence between molecular structure and polymerizability observed for LB films may not be directly applicable for predicting the properties of these surface attached diacetylene monolayers. In this report the influence of changes in the molecular architecture on the structure of the SAMs and their impact on the monolayer structure and the polymerization process are examined using various photopolymerizable diacetylenic disulfides:

$$\begin{array}{l} CH_3\text{-}(CH_2)_n\text{-}C{\equiv}C\text{-}C{\equiv}C\text{-}(CH_2)_m\text{-}S \\ \hphantom{CH_3\text{-}(CH_2)_n\text{-}C{\equiv}C\text{-}C{\equiv}C\text{-}(CH_2)_m\text{-}}| \\ CH_3\text{-}(CH_2)_n\text{-}C{\equiv}C\text{-}C{\equiv}C\text{-}(CH_2)_m\text{-}S \end{array}$$

The architecture of the monomolecular assemblies was adjusted synthetically by varying both the tail (n = 7,11,15) and spacer (m = 4,6,9) length.[28] A n,m-DA and n,m-PDA designation is used to identify the nonpolymerized and the polymerized monolayer architectures, respectively. The overall film quality was investigated employing contact angle measurements.[25] The internal order of the monolayers was probed by use of grazing angle infrared spectroscopy[25,29] and the polymerization was monitored by resonance Raman spectroscopy.[25,30]

Results

Wettability

To investigate the interdependence of the molecular architecture and the polymerizability of diacetylene containing SAMs, several diacetylene containing disulfides were synthesized[28] and self-assembled from solution onto gold surfaces.[29] The overall quality of these films was investigated by contact angle measurements. Although a large lateral area is probed when a drop of solvent is placed on the monolayer, interactions probe only the outermost methylene region. As shown in Tab. 1, contact angles for water on the longest tail monolayers (15,9 PDA; 15,6 PDA; 15,4 PDA) are almost identical and in good agreement with monolayers formed from octadecane thiol (C18). In contrast, contact angles exhibit a significant diminuition as the length of the tail region is decreased (11,9 PDA, 7,9 PDA).[25] This result is likely due to an increase of the disorder in the short tail chain, exposing more methylene groups to the surface as has been observed for short alkanethiols.[31] The contact angle measurements indicate a high overall film quality, where the tail region influences the outer chain ordering.

Tab. 1: Advancing contact angles for water on polymerized self-assembled monolayers of diacetylenic disulfides on gold (± standard deviation of five measurements).

	15,4 PDA	15,6 PDA	15,9 PDA	11,9 PDA	7,9 PDA	C18
contact angle	110 ± 3,7	109 ± 3,9	109 ± 1,8	105 ± 3,4	96 ± 2,0	113 ± 2,2

Polymerization

UV/Vis spectroscopy is the method of choice to monitor the polymerization of diacetylene compounds since the conjugated polymer backbone shows a strong absorption in the visible region. However, this method is not applicable to SAMs on gold due to the overlap of the polymer absorption band with the plasmon bands of the substrate.[25] Resonance Raman spectroscopy provides a good alternative approach due to the selective measurement of the conjugated polymer backbone. Indeed the resonant enhancement of the Raman signal for the polymer backbone allows monitoring of the polymerization process without interference from nonresonant transitions. Alkyl-based monolayers and nonpolymerized diacetylenes do not show any signal under the conditions used in this study. Furthermore, photoselection in resonant Raman spectroscopy can be used to discriminate the blue and red polymer phases. Excitation at 633 nm is expected to be in resonance with the blue phase polymer only, while excitation at 532 nm probes both blue and red phase polymer.[7,25,30,32]

The spectra obtained for the SAMs at two excitation wavelengths (633 nm probing blue phase polymer and 532 nm probing both blue and red phase polymer) are shown in Figure 2. The transitions are assigned to vibrations in the double (1450 -1500 cm^{-1}) and triple (2080 - 2100 cm^{-1}) bond regions. As expected for a highly conjugated system, these transitions are considerably shifted relative to the transitions anticipated for the isolated bonds (1620 and 2260 cm^{-1}, respectively.[33] The transition in the 700 cm^{-1} region has not been conclusively assigned in the literature and is conjectured to arise from the in-plane stretching vibrations of the backbone.[7,21] Relative intensities of alkene and alkyne vibrational transitions at a given excitation wavelength are expected to correspond with the degree of polymerization into the specific form. However, intensity comparison between spectra obtained with differing excitation wavelengths is not feasible due to different experimental configurations.

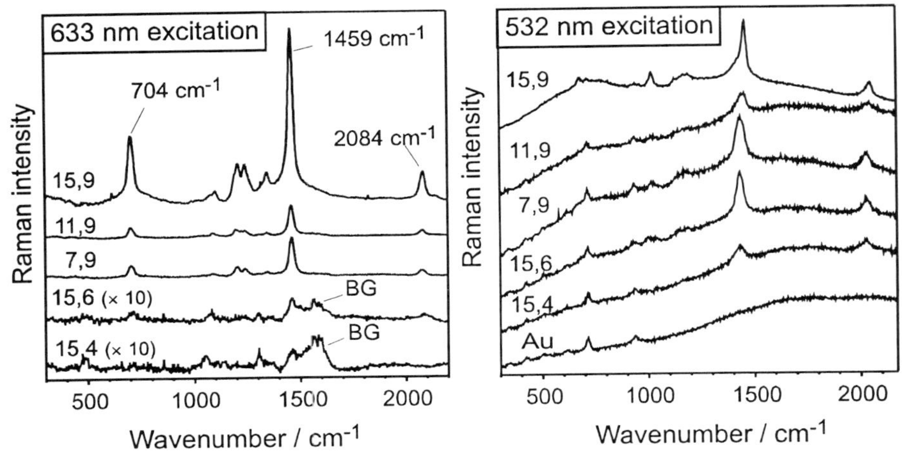

Figure 2: Resonance Raman spectra for two excitation wavelengths as function of molecular architecture (different scaling, see text).

The spectra indicate that for all molecular architectures polymerization can be observed, however the polymerization efficiency is clearly different. For 15,9 PDA there is a significant concentration of blue phase polymer, but for 11,9 PDA and 7,9 PDA lower intensities in the spectra for 633 nm excitation indicate less blue phase polymer. As the spacer length is decreased (15,9 PDA → 15,6 PDA → 15,4 PDA) the blue form decreases markedly until no signal is observed for the shortest spacer. However, excitation at 532 nm, which probes both red and blue phase polymer, shows the presence of polymer for all molecular architectures. For the shortest spacer the signal is small, indicating that the 15,4 PDA has only a small degree of polymerization.[25)]

Alkyl chain crystallinity

In LB film studies, decreasing alkyl chain crystallinity has been correlated with decreasing conjugation length, leading to the red-phase polymer. This correlation may arise from a decrease in the polymer length caused by a misalignment of neighboring diacetylenes with more disordered alkyl chains. Equally likely, disordered alkyl chains place strain on the polymer backbone, decreasing the conjugation length. The alkyl chain crystallinity can be probed by infrared spectroscopy. A highly crystalline, all-trans configuration is expected to exhibit the asymmetric (ν_a-CH_2) and symmetric (ν_s-CH_2) methylene stretching transitions at 2919 cm^{-1} and 2850 cm^{-1}, respectively.[34)] The transition frequency for both transitions will be shifted to higher wavelengths for decreasing order. As can be seen in Figure 3, the spectra of the monolayers vary considerably. The IR spectra indicate some disorder for the shorter tail structures (11,9

DA and 7,9 DA), while the longest tail structures of 15,9, 15,6 and 15,4 show peak positions indicating a highly ordered structure (see Tab. 2).

Tab. 2: Band positions for the methylene stretch vibrations in monomeric and polymeric diacetylene SAMs

	ν_a-CH_2		ν_s-CH_2	
	DA	PDA	DA	PDA
15,4	2919	2919	2851	2851
15,6	2919	2919	2851	2851
15,9	2919	2919	2852	2851
11,9	2924	2924	2855	2855
7,9	2925	2926	2855	2855
octadecane thiol	2920		2851	

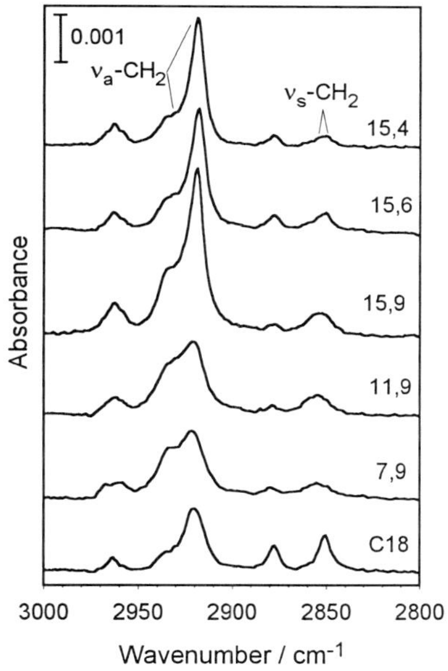

Figure 3: Fourier transform infrared spectra of the CH stretching region for diacetylene SAMs with various molecular architectures obtained at grazing incidence

However, upon increasing the spacer length from four to nine methylene units, an increasing transition appears as a higher frequency shoulder on the ν_a-CH_2 and ν_s-CH_2 peaks. The position of the transition is consistent with a non-ordered polymethylene chain. This transition is ascribed to an increased disordered component within the alkyl chains. The nonhomogeneity may arise from a domain structure with disordered chains at the domain boundaries. However, the high intensity of the disordered component in the IR spectra especially for the 15,9 DA monolayer is not consistent with electrochemical experiments, which indicate a very low defect density due to domain boundaries.[25,29] Alternatively, a model where the spacer and the tail have different methylene chain structures has been proposed.[29] In this model a tightly packed tail region provides a methyl surface with high blocking capability, masking a more loosely packed spacer region. It is important to note that a decreased crystallinity, as sensed by the IR-measurements for the spacer region, is not associated solely with the absence of order. It would also be observed if the spacer is in an arrangement with reduced chain-chain interactions, a so-called "conformationally ordered liquid state".

Discussion

Monitoring the polymerization of diacetylene SAMs with two excitation wavelengths revealed that the polymerization efficiency clearly depends on the molecular architecture of the compounds self- assembled onto the gold surface. High concentrations of blue phase polymer are found for the 15,9 PDA. Upon decreasing the length of the alkyl tail i.e. for 11,9 PDA and 7,9 PDA, the concentration of blue phase polymer is reduced significantly. Both the wettability and IR studies, which indicate more disorder in the SAMs with decreasing tail length, are consistent with disorder as origin of the limited polymerization efficiency. This observation is in accordance with LB film studies, which have shown that a high degree of order is necessary in the side chain region to obtain good polymerization efficiency.[8,11)] However, decreasing order cannot explain the significantly less efficient polymerization as the spacer length is decreased. Monolayers of 15,6 PDA and 15,4 PDA exhibit only little or no signal, when blue phase polymer is probed, although wettability and IR spectra indicate highly ordered alkyl chains in both cases. Indeed, the disordered portion, which manifests itself as a high frequency shoulder, is decreasing when the spacer becomes shorter. Since polymerizability is best for a long spacer, which shows a significant disordered portion, it can be conjectured that the spacer acts primarily to accomodate the alignment of adjacent diacetylene moieties. The longer spacer chain offers increased degrees of freedom and reduces the lattice strain from the misfit between the arrangement of the anchor groups at the gold surface and the packing of the diacetylene moieties. However in this case odd and even spacers are also expected to result in slightly different arrangements of the diacetylene units. For simple alkyl thiols it has been demonstrated that end group orientation is depending on the number of the methylene units.[35] Although a two-dimensional representation is always inadequate to give a clear picture of the SAM structure, the structure of diacetylene containing alkyl chains on gold as depicted in Figure 4 readily shows differences for odd and even numbered spacers. The structures have been drawn according to the knowledge about the structure of alkyl thiols on gold.[31, 36)] The distance of the sulfur atoms is 4.99 Å.[31)] The tilt angle of the alkyl chains θ depicted here, is slightly smaller than the value determined for pure alkylthiols (θ ~ 30°-40°).[31,34)] However, the tilt of the alkyl chain is a convolution of the tilt in the CCC-plane (paper plane) and perpendicular to the CCC-plane. There is experimental evidence that in the case of the diacetylene containing monolayers the tilt perpendicular to the CCC-plane is more important than for pure alkyl thiols[29)] and therefore a smaller tilt in the CCC-plane, as drawn here, is reasonable. Furthermore, there is experimental evidence that the diacetylene moieties form an internal scaffolding.[29,37)] The arrangement of the diacetylene moieties therefore is assumed to be very similiar to that found in crystals (i.e. $\gamma \sim 45°$). The major difference between the structures for even and odd numbered spacers, respectively, is the difference in the angle α (gold-sulfur-C). This angle should be 109° for a sp^3-hybridizised sulfur.[36)] While α is close to 109° for even numbered spacers, odd numbered all-trans spacers would result in an α of almost 180°. Therefore, some gauche bonds have to be in the odd numbered spacer chains and a higher level of disorder may be expected.

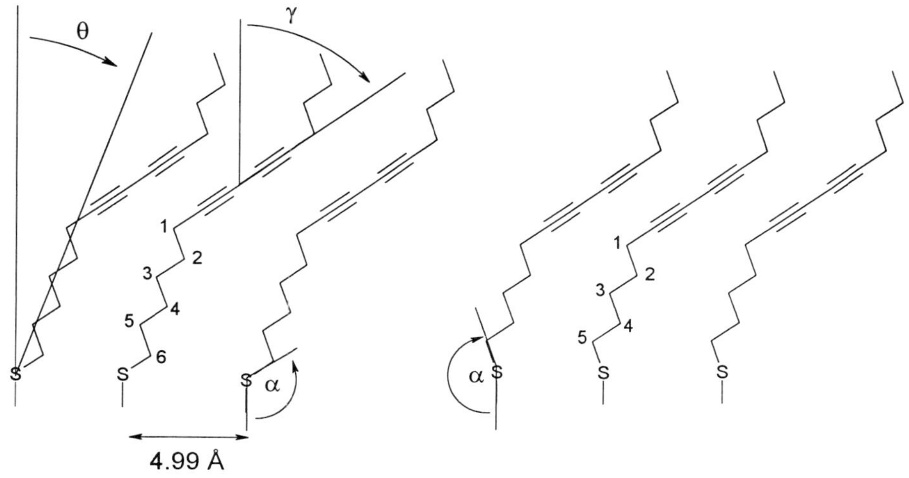

Figure 4: Schematic representation of the arrangement in a diacetylene containing monolayer for an even (6 methylene units, left) and odd (five methylene units, right) numbered spacer.

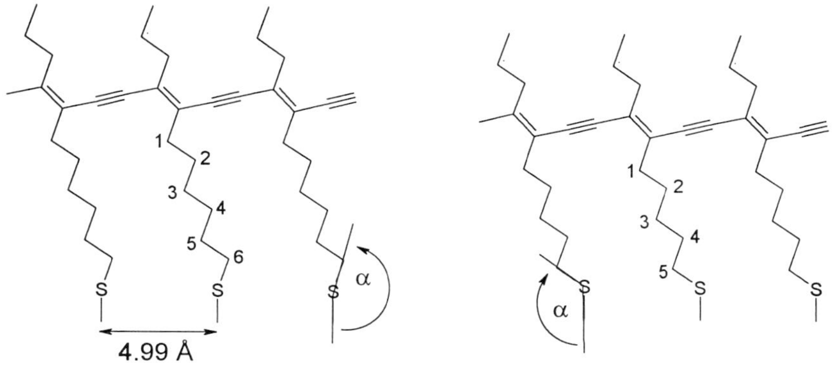

Figure 5: Schematic representation of the arrangement in a polydiacetylene monolayer for an even (6 methylene units, left) and odd (five methylene units, right) numbered spacer.

Although a packing favorable for the polymerization can be adapted in the case of even numbered spacers without gauche bonds, this does not necessarily imply that the polymerization is more favorable than in the case of odd numbered spacers. Because of the hybridization change in the diacetylene moiety upon polymerization, some structural changes have to take place within the spacer region. Figure 5 depicts a two-dimensional representation of the structure in a polymerized SAM. Again, the major difference between the structures for even and odd numbered spacers, respectively, is the difference in the angle α (gold-sulfur-C). However, upon polymerization and the concomitant change of the hybridization, an odd numbered spacer becomes more favorable. In polydiacetylene SAMs α is close to 109° for odd numbered spacers and would be almost 180° for even numbered spacers. This might cause too much strain to obtain the highly conjugated blue phase polymer in the case of even numbered spacers, since the anchor group cannot change its orientation as is possible in LB films.[3)] In particular too much strain can be anticipated for spacers too short to accomodate both reasonable bond angle for the anchor group and the change in the hybrization at the polymer backbone, even with gauche conformations. Thus, a decreased polymerization efficiency for diacetylene SAMs with short even numbered spacers can be expected. The increased disordered portion in the diacetylene SAMs with an odd numbered (and longer) spacer as well as the decreased polymerization efficiency for even numbered spacers, are in fact experimentally observed.

Conclusion

The influence of the molecular architecture of diacetylene containing disulfides on the structure and polymerizability of self-assembled monolayers on gold was investigated. The position of the diacetylene group was varied systematically by adjusting the length of the tail and spacer. The length of the tail group determines the order within the system. The longer the tail the higher the crystallinity of the SAM and the more efficient the polymerization. The spacer mainly acts to accomodate an arrangement of the diacetylene units favorable for polymerization. It can be conjectured that the arrangment is influenced by the length of the spacer and the number of the methylene units (odd or even), since one end of the spacer is tethered to the surface. An even numbered spacer can accomodate an arrangement of the diacetylene units without the need to distort the all-trans configuration, while some gauche bonds are neccessary for an odd numbered spacer. However, upon polymerization and the concomitant change of the hybridization, an odd numbered spacer is more favorable. So the molecular architecture of the diacetylene compound not only has to allow an arrangement of the triple bonds favorable for polymerization, but also to accomodate both the change in hybrization and a reasonable bond angle at the anchor group. To substantiate the conclusion drawn here, additional architectures have to be investigated and a more sophisticated modeling of the monolayer structure has to be applied to the system.

References

[1)] G. Wegner, *Z. Naturforsch. B* **24**, 824 (1969)

[2)] B. Tieke, G. Wegner, D. Naegle, H. Ringsdorf, *Angew. Chem. Int. Ed. Engl.* **15**, 764 (1976)

3) B. Tieke, H.J. Graf, G. Wegner, B. Naegele, H. Ringsdorf, A. Bauerjie, D. Day, J.B. Lando, *Colloid Polym. Sci.* **255**, 521 (1977)

4) G. Lieser, B. Tieke, G. Wegner, *Thin Solid Films* **68**, 77 (1980)

5) D. Day, J.B. Lando, *Macromolecules* **13**, 1478 (1980)

6) D. Day, J.B. Lando, *Macromolecules* **13**, 1483 (1980)

7) B. Tieke, D. Bloor, R.J. Young, *J. Mater. Sci. (London)* ***17***, 1156 (1982)

8) B. Tieke, G. Lieser, *J. Colloid Interface Sci.* **88**, 471 (1982)

9) B. Tieke, G. Lieser, K. Weiss, *Thin Solid Films* **99**, 95 (1983)

10) V. Enkelmann, *Adv. Polym. Sci.* **63**, 91 (1984)

11) B. Tieke, *Adv. Polym. Sci.* **71**, 79 (1985)

12) J.B. Lando, in: *"Polydiacetylenes"* D. Bloor, R.R. Chance Eds., Nijhoff, Dordrecht, The Netherlands (1985), pp. 363

13) M. Schott, G. Wegner, in: *"Nonlinear optical properties of organic molecules and crystals"*, Eds. D.S. Chemla, J. Zyss, Academic Press, Orlando (1987)

14) G. Cao, T.E. Mallouk, *J. Solid State Chem.* **94**, 59 (1991)

15) N. Mino, H. Tamura, K. Ogawa, *Langmuir* 7, 2336 (1991)

16) K. Kuriyama, H. Kikuchi, T. Kajiyama, *Langmuir* **12**, 2283 (1996)

17) D.H. Charych, J.O. Nagy, W. Spevak, M.D. Bedvarski, *Science* **261**, 585 (1993)

18) Q. Cheng, R. Stevens, *Adv. Mater.* **9**, 481 (1997)

19) A. Reichert, J.O. Nagy, W. Spevak, D. Charych, *J. Am. Chem. Soc.* **117**, 829 (1995)

20) J.J. Pan, D. Charych, *Langmuir* **13**, 1365 (1997)

21) D.N. Batchelder, S.D. Evans, T.L. Freeman, L. Häussling, H. Ringsdorf, H. Wolf, *J. Am. Chem. Soc.* **116**, 1050 (1994)

22) T. Kim, Q. Ye, L. Sun, K.C. Chan, R.M. Crooks, *Langmuir* **12**, 6065 (1996)

23) K.C. Chan, T. Kim, J.K. Schoer, R.M. Crooks, *J. Am. Chem. Soc.* **117**, 5875 (1995)

24) M.D. Mowery, C.E. Evans, *J. Phys. Chem. B* **101**, 8513 (1997)

25) H. Menzel, M.D. Mowery, M. Cai, C.E. Evans, *J. Phys. Chem. B* **102**, 9550 (1998)

26) T. Kim, K.C. Chan, R.M. Crooks, *J. Am. Chem. Soc.* **119**, 189 (1997)

27) H. Menzel, M.D. Mowery, M. Cai, C.E. Evans, *Adv. Mater.* in press (1999)

28) M.D. Mowery, C.E. Evans, *Tetrahedron Lett.* **38**, 11 (1997)

29) H. Menzel, M.D. Mowery, M. Cai, C.E. Evans, *Macromolecules* submitted

30) M. Cai, M.D. Mowery, H. Menzel, C.E. Evans, *Langmuir* submitted

31) A. Ulman, *"An Introduction to ultrathin Organic Films - From Langmuir-Blodgett to Self-Assembly"* Academic Press, San Diego (1991)

32) A.J. Campbell, C.K.L. Davies, D.N. Batchelder, *Macromol. Chem. Phys.* **199**, 109 (1998)

33) D. Lin-Vien, N.B. Colthup, W.G. Fateley, J.G. Grasselli in: *"The Handbook of Infrared and Raman Characteristic Frequencies of Organic Molecules"*, Academic Press, San Diego (1991)

34) M.D. Porter, T.B. Bright, D.L. Allara, C.E.D. Chidsey, *J. Am. Chem. Soc.* **109**, 3559 (1987)

35) P.E. Laibinis, G.M. Whitesides, D.L. Allara, Yu-Tai Tao, A.N. Parikh, R.G. Nuzzo, *J. Am. Chem. Soc.* **113**, 7152 (1991)

36) H. Sellers, A. Ulman, Y. Shnidman, J.E. Eilers, *J. Am. Chem. Soc.* **115**, 9389 (1993)

37) M.D. Mowery, H. Menzel, Mei Cai, C.E. Evans, *Langmuir* **14**, 5594 (1998)

Synthesis and Second Harmonic Generation Studies of NLO-functionalized Polyimides

Hong-Ku Shim*, Chong-Bok Yoon, Han Young Woo

Department of Chemistry, Korea Advanced Institute of Science and Technology, Taejon 305-701 Korea

SUMMARY: Several NLO-functionalized polyimides were synthesized through novel synthetic pathways and their NLO properties were characterized. For preventing the disadvantage in poly(amic acid) precursor route, the Mitsunobu reaction and alternative synthetic route for NLO polyimides were chosen. The second-order NLO coefficients ($\chi^{(2)}$) of the synthesized polyimides were between 48pm/V and 122pm/V. The thermal stability of nonlinearity characterized showed that all polyimide films retain their NLO activities at temperatures below 100 °C.

Introduction

Due to the mechanical and cost advantages over inorganic counterpart, organic polymers have been extensively researched as the material for electro-optic switch, frequency doubler and waveguides.[1,2] However, the relaxation of NLO activity of polymeric film at high temperature was severe problem for actual application.[3,4] For solving such problem, many attempts were introduced including crosslinking after alignment of dipole[5,6] and/or rendering polymer backbone rigid.[7-9] Among the latter, polyimide derivatives were mostly studied because of their excellent thermal stabilities and high glass transition temperatures (T_g).[10,11] Traditional synthetic pathway of polyimides involved the reaction of diamine and dianhydride monomers, resulting the poly(amic acid) precursors which lead to polyimide structure by chemical or thermal treatment.[12] This preparation method, however, has restriction in some monomers with NLO chromophores because it requires chemically or thermally harsh condition during imidization process[10] or any diamines synthesis.

In this study, we synthesized polyimides via somewhat different pathway. First, for avoiding such harsh conditions we synthesized polyimides directly via Mitsunobu reaction.[13] Since Mitsunobu reaction provides relatively mild condition, labile NLO chromophores could be introduced in polyimides more easily. Second, for preventing any hard condition to take place during the synthesis of NLO-phore-bearing diamine monomers, we introduced NLO chromophore into the dianhydride monomers, which enable much wider choice of NLO

 CCC 1022-1360/99/$ 17.50+.50/0

chromophores. In these regards, we report the synthesis and second-harmonic generation properties of the polyimides.

Result and discussion

The Synthesis of Polyimide

For avoiding chemically and thermally harsh process, we introduced two synthetic methods involving the preparation of polyimides. One is the direct synthesis via Mitsunobu reaction (Scheme 1) and the other is related to the synthesis of chromophore-containing dianhydride monomers. (Scheme 2,3)

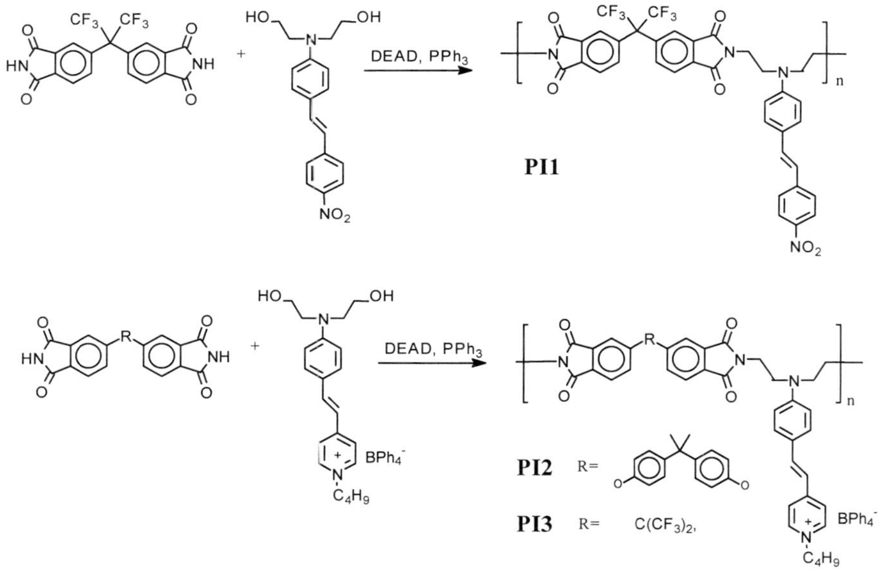

Scheme 1. Synthesis of polyimide via Mitsunobu reaction

The first synthetic procedures for polymers are shown in Scheme 1. The diimide and dihydroxy monomers were synthesized as reported earlier.[14,15)] The polymerization reaction was executed by Mitsunobu reaction using diethyl azodicarboxylate (DEAD) and triphenylphosphine in anhydrous tetrahydrofuran (THF) solvent to give soluble polyimides directly. By performing direct polymerization under the mild Mitsunobu condition, the harsh imidization process of the poly(amic acid) can be avoided and the synthesis of chromophore-containing diamine monomers is also unnecessary. Therefore, the chromophore which has a large first-order hyperpolarizability could be incorporated in the polyimide backbone without decomposition of the chromophore molecules. We chose hemicyanine dye as the NLO component since it has extraordinarily high β value compared with other NLO chromophores,[16] but is relatively labile toward chemical or thermal harshness.[17]

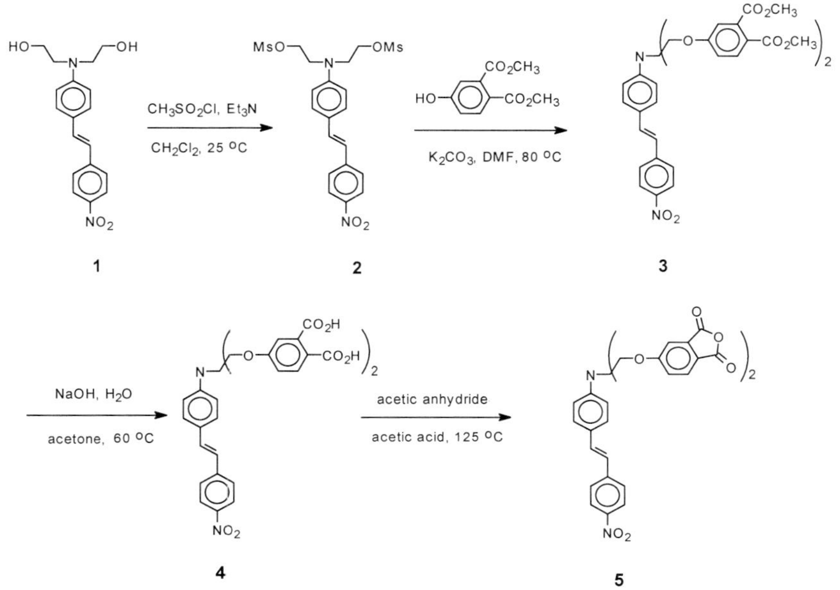

Scheme 2. Synthetic route for the NLO chromophore-containing dianhydride monomer.

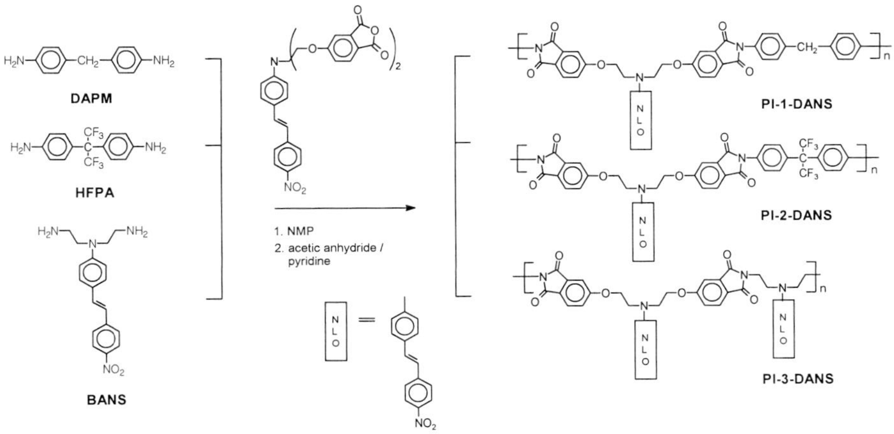

Scheme 3. Synthetic route for the polyimides.

The synthetic routes for the chromophore-containing dianhydride monomer and resulting polyimides are presented in Scheme 2 and 3. We incorporated 4-bis(hydroxyethyl)amino-4'-nitrostilbene (compound **1**), which is representative for second-order nonlinear optical material, into the polyimide backbone. It was synthesized according to the method described in the previous literature.[18] This simple synthetic route may be applied to many kinds of NLO chromophores which have dihydroxy functional groups. The difficulty in the synthesis of chromophore-containing diamine compounds has limited the type of chromophores which may be incorporated into the polyimide backbone. However, thorough this simple methodology, a wider variety of chromophores can be incorporated into the polyimide backbone without the synthesis of NLO-diamine compounds. To vary the physicochemical and optical properties such as solubility, glass transition, second-order optical nonlinearity, *etc.*, three types of diamine compounds were used to synthesize final polymers. Both of PI1- and PI2-DANS are different with PI3-DANS in the chromophore contents in the polymer backbone. Until now, in most NLO polymers synthesized previously, only one monomer partner contains a NLO chromophore because of solubility, synthetic difficulty, optical quality, *etc.*, of final polymers. There have been very few reports where the functionalities are introduced into both comonomers.[19] However, in the PI3-DANS case, both the monomer partners have NLO chromophores. This induces double chromophore density different from those of PI1-, PI2-DANS cases, from which much higher NLO activity can be anticipated. In addition, due to presence of alkyl space tether groups and ether linkages it is very flexible to give good processibility also.

The Physiochemical and Optical Properties of Synthesized Polyimides

The molecular weights of the polyimides synthesized through Mitsunobu reaction(Tab. 1), determined by gel permeation chromatography (GPC) with a polystyrene standard, ranged in 8,200 ~ 14,400 for weight averaged molecular weight (M_w). The PI2 and PI3 showed lower molecular weight than that of PI1. The bulkiness and ionic character of tetraphenylborate anion should have retarded the Mitsunobu reaction to give shorter polymer chain. The weight averaged molecular weights of three polymers synthesized from NLO-functionalized dianhydride monomer, determined by GPC with a polystyrene standard, were 15,900 (M_w/M_n=1.59) for PI1-DANS, 38,300 (M_w/M_n=2.28) for PI2-DANS, and 25,300 (M_w/M_n=2.13) for PI3-DANS, respectively.

These polymers were readily soluble in THF, cyclohexanone, *etc.* From cyclohexanone solutions of PI1, PI2 and PI3, good optical quality thin films could be prepared onto glass substrate by spin coating. The polymer structures were characterized by UV-visible absorption and FT-IR spectroscopy. The FT-IR transmittance spectra of the polymers showed structural similarity concerning with the formation of imide group in all cases. The peaks at 1772 and 1719 cm^{-1} are ascribed to the carbonyl groups in the imide linkages. These results are well consistent with the proposed structures, indicating that the NLO chromophore survived the imidization process. In the UV absorption spectra, the polyimides containing DANS as NLO chromophore (PI1, PI1-DANS, PI2-DANS, PI3-DANS) showed absorption maximum around 430 nm corresponding to that of dialkylaminonitrostilbene (DANS) moiety. The PI2 and PI3 also showed similar characteristics of hemicyanine dye that gave maximum absorption at 467 nm. No detectable change of UV absorption in PI2 and PI3 indicated that the chain polarity of polymer backbones of PI2 and PI3 was not significantly different from each other.

Tab. 1. The Properties of polyimides synthesized via Mitsunobu reaction

Polymer	M_w(PDI)	T_g(°C)	T_d(°C) [a)]	α [b)] (μm^{-1})	$\chi^{(2)}_{33}$ (pm/V)	λ_{max} (nm)	$\chi^{(2)}_{33}(0)$ [c)] (pm/V)
PI1	14,400(1.87)	198	332	0.33	82	432	26
PI2	10,800(1.48)	135	226	0.81	-	467	-
PI3	8,200(1.89)	175	227	0.82	122	467	24

a) temperature of 95% weight retention.
b) absorption constant
c) $\chi^{(2)}_{33}(0)$ is calculated with two-level model

Tab. 2. The Properties of of polyimides from NLO-functionalized dianhydride monomer

Polymer	M_w (PDI)	T_g (°C)	T_d (°C)	α (μm^{-1})	$\chi^{(2)}_{33}$ (pm/V)	λ_{max} (nm)	$\chi^{(2)}_{33}(0)$ (pm/V)
PI1-DANS	15,900 (1.59)	161	318	0.49	74	439	20
PI2-DANS	38,300 (2.28)	181	315	0.33	48	431	14
PI3-DANS	25,300 (2.13)	168	305	0.57	72	438	20

Thermal properties of the polyimides were examined by TGA and DSC measurements. In the TGA thermograms of polyimides, the thermal stabilities of those polymers were observed.

Since the polyimide chain maintained its stability over 500°C,[20] the weight losses should solely result from the chromophore decomposition. We defined the T_d as the temperature of 95% weight retention. In Tab.1, the T_d of PI1 was 332 °C, that is similar result with NLO-functionalized polyimide prepared by other methods. The T_d's of PI2 and PI3, however, were around 226°C showing relatively unstable thermal properties. The instability of PI2 and PI3 is attributed to the hemicyanine moiety with bulky tetraphenylborate anion.[15] In the comparison of PI1 with PI3, the PI1 showed higher T_g than PI3, which can be thought of as the effect of chromophore difference. The bulkier hemicyanine dye should lower the crystallinity to yield the glass transition at lower temperature. Especially low T_g of PI2 is thought to be due to more flexible units (bisphenol A unit). In the case of PI1-, PI2-, PI3-DANS, it is noticed that the initial decay of the system takes place in *ca.* 305-318 °C range, which is ascribable for the decomposition of the stilbene moiety in the NLO-phore, and the breakdown of the polymer main chain occurred at around 455 °C. The DSC curves showed the glass transition area starting at 161 °C for PI1-DANS, 181 °C for PI2-DANS, and 168 °C for PI3-DANS. These are relatively low temperatures compared to those of common rigid polyimdes and are attributed to a flexibility of polymer backbone containing ether linkages and alkyl tether groups. In the PI2-DANS case, the CF_3- unit in the polymer backbone increased the chain rigidity due to the polarity of trifluoromethyl group and thus resulted in high T_g.

The Sample Preparation and Measurement of Second-order NLO Properties

The second-order NLO process requires that the material must be noncentrosymmetric. For the amorphous polyimide films, the centrosymmetry can be broken by corona-discharge poling when the film is heated up to the T_g so that the molecular dipole can be oriented.[5,21] The PI1 and PI3 were dissolved in cyclohexanone and cast onto indium-tin oxide (ITO) surfaces to form thin films of 1-3μm. As the decomposition temperatures of polyimides were higher than T_g of those polymers, we performed the poling step at slightly higher temperature than T_g of each polymer in electric field under nitrogen flow. Tungsten wire of 20μm thickness was used to generate the corona charge. The second-order nonlinearity ($\chi^{(2)}_{33}$)of the polymer samples (PI1, PI3) was measured with second harmonic generation (SHG) method at 1064nm, as reference wavelength. The $\chi^{(2)}_{33}$ values of PI1 and PI3 measured by the angular dependence method using a quartz crystal as the reference[22,23] were found to be 82 and 122pm/V, respectively. The PI3 which contains the hemicyanine dye showed larger $\chi^{(2)}_{33}$ value compared with that of PI1 containing DANS moiety. However, the resonance effect on

SHG measurement should not be ignored in real device applications. We calculated nonresonant values by two-level model.[24)] The calculated $\chi^{(2)}_{33}$ (0) value of PI1 became similar to that of PI3. Although hemicyanine dye has larger β value than that of DANS to increase $\chi^{(2)}$ values, its bulkiness should have lowered chromophore density in the polymer films to lower $\chi^{(2)}_{33}(0)$ values.[4] In the case of PI1-, PI2-, PI3-DANS, the results of second harmonic generation measurements performed at a wavelength of 1064 nm, revealed large $\chi^{(2)}_{33}$ values, 74 pm/V for PI1-DANS and 72 pm/V for PI3-DANS. (Fig. 1). It was found that the disperseless $\chi^{(2)}_{33}$ value of PI1-DANS was *ca.* 20 pm/V and *ca.* 14 pm/V for PI2-DANS and *ca.* 20 pm/V for PI3-DANS respectively. The relatively low value of PI3-DANS in comparison to others, irrelevant of the double chromophore density, may result from the ineffective poling due to the matrix rigidity.

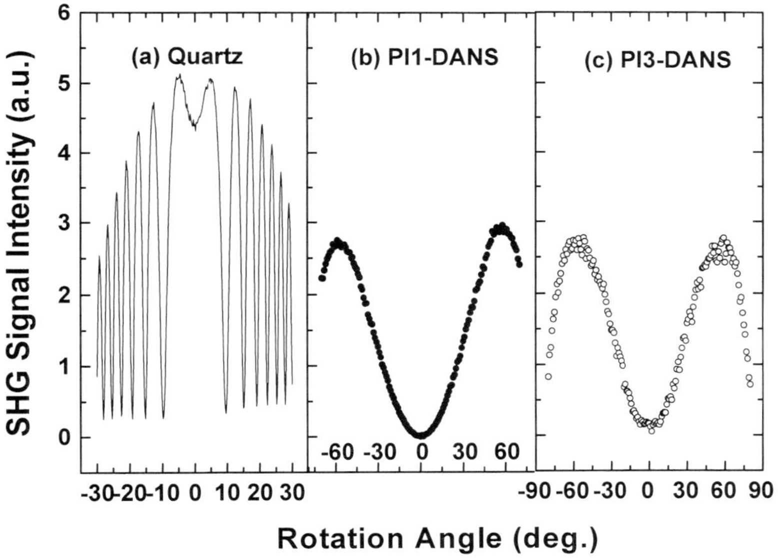

Fig.1: Maker-fringe pattern of (a) quartz crystal, (b) PI1-DANS and (c) PI3-DANS

As the thermal relaxation of dipole orientation is related to the free volume that is concerned to the glass transition temperature, the thermal stability of the NLO properties of the poled polymers can be predicted from the T_g of each polymer. At room temperature, all polyimides were significantly stable and there is no decay in the intensity of SHG signal. All polymers sustained their nonlinearity at 100°C in air for several hundred hours. Because the observation of temporal stability is a time-consuming experiment, we performed the temperature-dependent SHG measurement. Fig. 2 shows the dynamic thermal stabilities of PI1 and PI3. The PI1 and PI3 maintained 90% of original nonlinearity at temperatures below 150°C. The SHG signal of both polyimides was quite stable until the temperature reached 150°C. As the temperature reached the glass transition temperature of the polymers, fast relaxation of NLO activity was observed. The PI3 showed sudden relaxation around 170°C corresponding to the T_g of PI3. In the case of PI1, it showed similar relaxation profile with that of PI3, although it has higher T_g (198°C). We thought that the bulkiness of NLO chromophore should affect the thermal stability of NLO activity. DANS chromophore of PI1 is less bulky than the hemicyanine dye containing tetraphenylborate anion. Bulkier tetraphenylborate ion should prevent the aligned dipoles (hemicyanine chromophore) from reorienting to random direction.[14] So PI1 and PI3 showed similar thermal stability, even though PI3 has lower T_g.

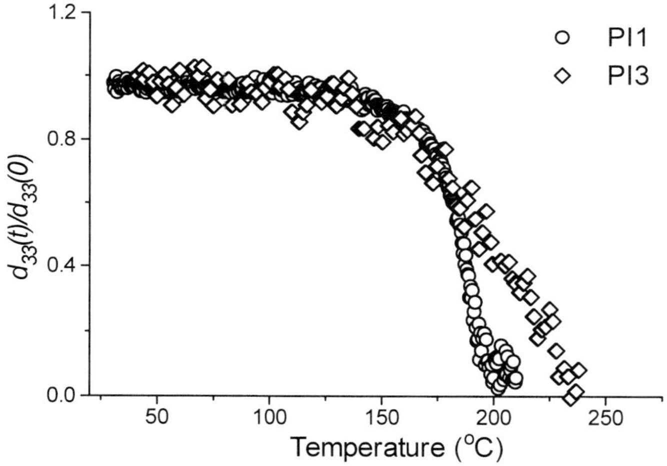

Fig. 2: Dynamic thermal stability of NLO activity of PI1 and PI3 (temperature rate: 10°C/min, in air).

Conclusion

We successfully synthesized the NLO-functionalized polyimides containing DANS and the hemicyanine dye. Mild Mitsunobu reaction condition enables the incorporation of chemically labile hemicyanine dye into polyimide backbone. These polymers showed high nonlinearity and good temporal stability. The Mitsunobu reaction is a versatile way to the synthesis of polyimide which contains the NLO chromophore of high β value. We also prepared three kind of polyimides through a new alternative synthetic route in which NLO chromophore was incorporated into the dianhydride monomer. They showed high second-order optical nonlinearity and good thermal stabilities of the aligned dipoles. The measured $\chi^{(2)}_{33}$ values ranged between 48 and 122 pm/V. The proposed synthetic route may be applied to various NLO chromophores and due to the versatility of the reaction scheme, many new materials can be expected.

Acknowledgement

We are gratefully acknowledged to Dr. Jin (KRICT), Mr. Mun and Prof. Yoon (Dept. of Physics, KAIST) for measuring the $\chi^{(2)}$ values of the polymer films. This work was supported by the the Center for Advanced Functional Polymers through KOSEF.

References

1) P. N. Prasad, D. J. Williams, "*Introduction of Nonlinear Optical Effect in Molecules and Polymers*", Wiley, New York 1991
2) Chemla, D. S. "*Nonlinear Optical Properties of Organic Materials and Crystals*", Academic Press, New York 1987
3) L. R. Dalton, A. W. Harper, R. Ghosn, W. H. Steier, M. Ziari, H. Fetterman, Y. Shi, R. V. Mustacich, A. K-Y. Jen, K. J. Shea, *Chem. Mater.*, 7, 1060 (1995)
4) D. M. Burland, R. D. Miller, C. A. Walsh, *Chem. Rev.*, **94**, 31 (1994)
5) C. B. Yoon, H. K. Shim, *Macromol. Chem. Phys.*, **199(11),** 2433 (1998)
6) C. K. Park, J. Zieba, C. F. Zhao, B. Swedek, W. M. K. P. Wijekoon, P. N. Prasad, *Macromolecules*, **28**, 3713 (1995)
7) C. B. Yoon, K. J. Moon, H. K. Shim, *Macromolecules*, **29(17),** 5754 (1996)
8) T. Verbiest, D. M. Burland, M. C. Jurich, V. Y. Lee, R. D. Miller, W. Volksen, *Science*, **268**, 1604 (1995)
9) R. Dagani, *Chem. Eng. News*, **74**, 22 (1996)
10) T. Verbiest, D. M. Burland, M. C. Jurich, V. Y. Lee, R. D. Miller, W. Volksen, *Macromolecules*, **28**, 3005 (1995)
11) T. A. Chen, A. K-Y. Jen, Y. Cai, *J. Am. Chem. Soc.*, **117**, 7295 (1995)
12) D. Yu, A. Gharavi, L. Yu, *J. Am. Chem. Soc.*, **117**, 11680 (1995)
13) O. Mitsunobu, *Synthesis*, **1981** *January*, 1.
14) C. B Yoon, K. J. Moon, H. K. Shim, K. S. Lee, *Mol. Cryst. Liq. Cryst.*, ***316***, 43 (1998)
15) K. J. Moon, H. K. Shim, K. S. Lee, J. Zieba, P. N. Prasad, *Macromolecules*, **29(3),** 861 (1996)
16) O. K. Kim, L. S. Choi, *Langmuir*, **10**, 2842 (1994)
17) C. B. Yoon, H. K. Shim, in preparation
18) S. Yang, Z. Peng, L. Yu, *Macromolecules*, **27**, 5858 (1994)
19) N. Nemoto, F. Miyata, Y. Nagase, *Chem. Mater*. **9**, 304 (1997)
20) S. H. Lin, F. M. Li, S. Z. D. Cheng, F. W. Harris, *Macromolecules*, **31(7)**, 2080 (1998)
21) T. J. Marks, M. A. Rather, *Angew. Chem., Int. Ed. Engl.* **34**, 155 (1995)
22) W. N. Herman, L. M. Hayden, *J. Opt. Soc. Am. B*, **12(3)**, 416 (1995)

[23] J. Jerphagnon, S. K. Kurtz, *J. Appl. Phys.*, **41**, 1667 (1970)
[24] M. Kauranen, T. Verbiest, C. Boutton, M. N. Teerenstra, K. Clays, A. J. Schouten, R. J. M. Nolte, A. Persoons, *Science*, **270**, 966 (1995)

Synthesis and Solution Properties of Rodlike Polyelectrolytes

Matthias Wittemann, Steffen Kelch, Jürgen Blaul, Peter Hickl, Birgit Guilleaume, Gisela Brodowski, Andreas Horvath, Matthias Ballauff*, Matthias Rehahn*

Polymer Institute, University of Karlsruhe, Kaiserstrasse 12, D-76128 Karlsruhe, Germany

SUMMARY: The synthesis and the solution behavior of rigid, rodlike cationic polyelectrolytes having (*i*) poly(*p*-phenylene) backbones and (*ii*) main chains composed of 4,4"-bis(2,2':6',2"-terpyridine)2',5'-dihexyl-*p*-terphenyl moieties and ruthenium(II) centers are presented. All these polymers are shown to have a homogeneous constitution and degrees of polymerization of up to $P_n \approx 70$. Their solution properties were analyzed using viscosimetry, small-angle X-ray scattering (SAXS) and osmometry. Pronounced polyelectrolyte effects were found in salt-free solutions using viscosimetry. Small-angle X-ray scattering demonstrated directly the strong correlation of the counterions and the macroions. The osmotic coefficients measured in salt-free solution as function of polyelectrolyte concentration were found to be lower than predicted by the cell model.

Introduction

The understanding of flexible polyelectrolytes in dilute solutions of low ionic strength still presents a considerable challenge in macromolecular science, despite of many decades of research[1-4]. This is due to the long-range nature of the coulombic forces between the charged macromolecules. In the case of flexible polyelectrolytes, a decrease of the ionic strength may lead to an expansion of the coils due to strong intramolecular forces as well as to stronger intermolecular electrostatic interactions. Conformationally rigid, rodlike polyelectrolytes, on the other hand, remain in their extended chain conformation regardless of the ionic strength of the system. Because conformative and excluded volume effects are ruled out here, only the intermolecular coulombic interactions determine the solution properties of these polymers.

Based on these considerations, a number of studies have been performed using naturally occurring rodlike helical polyelectrolyte systems such as DNA or xanthane[5]. However, at very low ionic strengths and at elevated temperatures the helical conformation and thus the rodlike shape are lost. Moreover, systematic variation of the charge density, *i.e.* the number of ionic groups per unit length, is not possible with the aid of these biopolymers. Therefore, the development of well-defined synthetic rodlike polyelectrolytes is necessary to analyze

 CCC 1022-1360/99/$ 17.50+.50/0

quantitatively intermolecular interactions, counterion condensation, and structure formation in solution depending on ionic strength, temperature, and polyelectrolyte concentration.

First synthetic rodlike polyelectrolytes were published in the early eighties[6,7] which based on poly(1,4-phenylenebenzobisoxazoles) and poly(1,4-phenylenebenzobisthiazoles). In the recent decade, we[8-12] and others[13,14] developed various efficient precursor routes to synthetic rodlike poly(*p*-phenylene) (PPP) polyelectrolytes which take advantage of both the concept of solubilizing side chains[15,16] and the efficient Pd-catalyzed aryl-aryl coupling reaction[17-19]. This progress was rendered possible mainly by (*i*) the high tolerance of the Pd-catalyzed polycondensation reactions toward functional groups in the starting materials and (*ii*) the outstanding thermal and chemical stability of the PPP backbone which allows transformation of the uncharged precursor PPPs into polyelectrolytes by a variety of organic reactions. By these precursor routes, high-molecular-weight carboxylated and sulfonated PPP polyelectrolytes with homogeneous constitution and well-known degrees of polymerization (P_n) have been prepared first (see, for example: Scheme 2, structures **7** and **8**). The polymers thus available combine exceptional hydrolytic, thermal and chemical stability with a high charge density of up to two ionic groups per *p*-phenylene repeating unit. Unfortunately, however, all these anionic PPP polyelectrolytes certainly dissolved in polar organic solvents or mixtures thereof with water but were insoluble in water or aqueous bases. Therefore, cationic PPP polyelectrolytes have been prepared as well which were expected to be much better soluble in water. The results thus obtained will be described in the first part of the present paper. Moreover, for comparison purposes, we additionally prepared rodlike cationic ruthenium(II) coordination polyelectrolytes. Hence, the second part of this paper presents a description of the synthetic route developed for the preparation of these readily soluble and constitutionally well-defined transition-metal containing macromolecules as well as of their properties in solution.

Synthesis of PPP polyelectrolytes

A series of uncharged precursor PPPs **3** - **5** has been prepared according to Scheme 1, bearing phenoxy substituents at the ends of either some or even all of their solubilizing *n*-hexyl side chains[10,11,20]. These phenoxy groups represent ideal precursor functionalities because they are inert under the conditions of the Pd-catalyzed polycondensation but can be activated easily and completely for the subsequent introduction of the desired charged groups via macromolecular substitution routes. The precursor strategy thus used for the preparation of the cationic PPP polyelectrolytes has moreover the distinct advantage that the uncharged precursors **3** - **5** can be fully characterized without being affected by the well-known problems associated with charged macromolecules. NMR spectroscopy proved their defect-free constitution, and number-average degrees of polymerization of up to $P_n = 70$ were determined reliably using membrane osmometry for all these precursors. The persistence length of the PPP backbone was estimated[21)] by Bohdanecky's method[22)] to be 22 nm. This is in good agreement with the results from a recent computer simulation[23)] but considerably higher than the value of 13 nm deduced from a similar PPP recently by Vanhee et al.[14)] Despite of this discrepancy, it seems to be justified to regard the PPP derivatives under consideration here (contour lengths of 10 to 40 nm) as rodlike entities.

Scheme 1

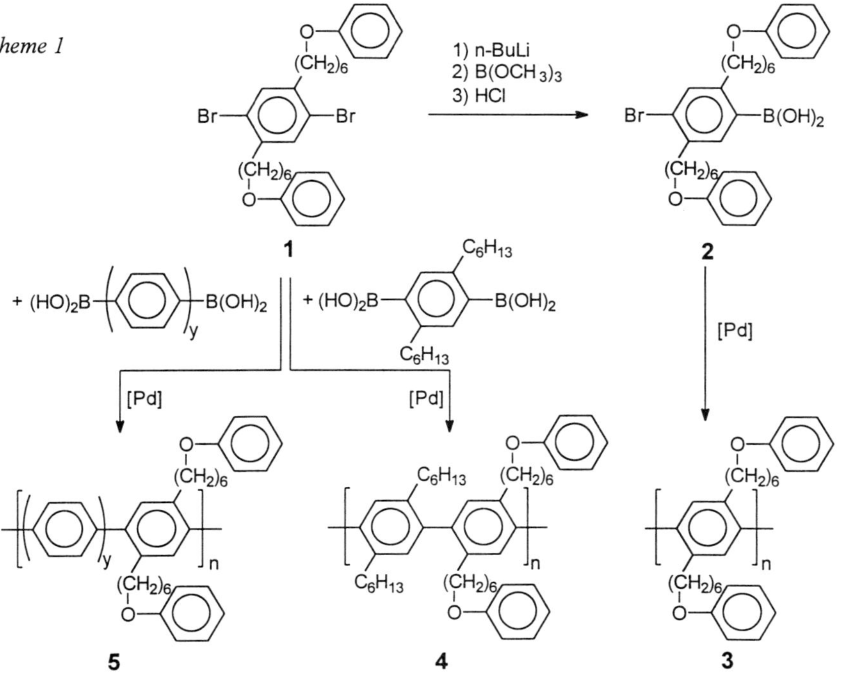

After their full characterization, the precursors **3** - **5** were converted into cationic PPP polyelectrolytes via the route shown in Scheme 2 for PPP derivative **3**: Firstly, the phenoxy groups were cleaved off quantitatively using trimethyliodosilane. In the second step, the cationic ammonium or pyridinium groups were introduced via quaternization of the 6-iodohexyl substituted precursor PPPs with tertiary amines such as triethylamine or pyridine. All the thus obtained PPP polyelectrolytes proved to be readily soluble in, for example, acetonitrile and dimethylacetamide, some of them even in pure water. Their ^{1}H and ^{13}C NMR spectra showed full conversion under the used conditions: The quaternization resulted in the complete disappearance of the iodomethylene absorptions, and all new signals in the NMR spectra were in full agreement with the molecular structures of the desired polyelectrolytes. Side reactions such as elimination or cross-linking processes were thus excluded for sure within the scope of accuracy of the NMR method. Moreover, no fractionation was admitted during the workup. Therefore, identical values of P_n can be assumed for the polyelectrolytes thus obtained as determined in advance for their precursors.

Scheme 2

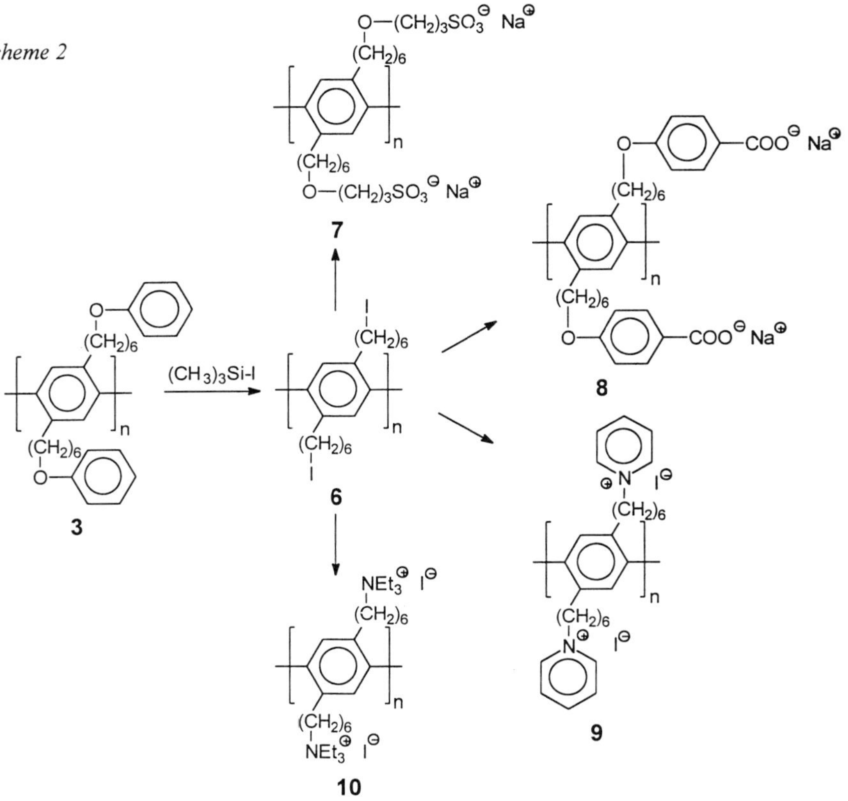

In addition to the readily water-soluble PPP polyelectrolytes such as **9** and **10** having two charged groups per phenylene repeating unit, we were also interested in making available further PPP derivatives having an even higher charge density. Therefore, an efficient two-step quaternization route was developed, leading to PPP polyelectrolytes **12** with as many as four tetraalkyl ammonium groups per *p*-phenylene repeating unit, *i.e.*, every 4.5 ångstrøms. For this purpose, precursor PPP **6** was first reacted with an approximately fourhundredfold molar excess of tetramethylethylenediamine (TMEDA) in a solvent mixture of chloroform and acetonitrile (Scheme 3). The large excess of TMEDA used in the synthesis of **11** is essential in order to suppress quaternization of both TMEDA nitrogen atoms in this first step which would lead to structural irregularities and crosslinks. After workup and purification by ultrafiltration, the remaining TMEDA amino groups of **11** were quaternized. For this purpose, the brownish, foamy material **11** was redissolved in dimethylsulfoxide and heated in the presence of a slight excess of iodoethane. The progress of the conversion was monitored using NMR spectroscopy. A prolonged reaction time and forced conditions were found to be necessary here to achieve complete conversion. The resulting polyelectrolyte solutions were again purified by ultrafiltration.

Scheme 3

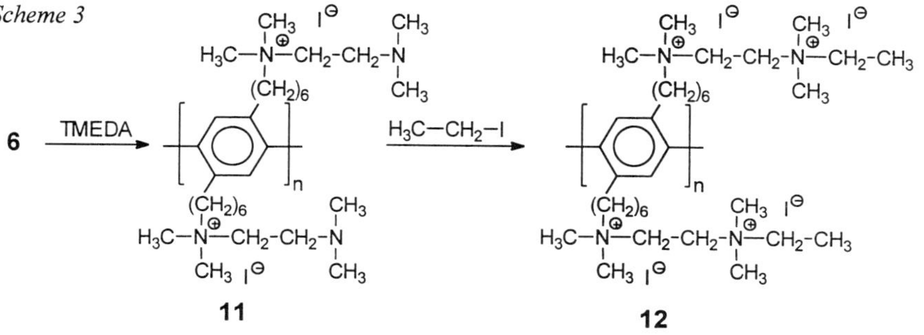

The homogeneous constitution of intermediate **11** as well as of the doubly quaternized polyelectrolyte **12** was proved using high-resolution NMR spectroscopy. Moreover, because no evidence of chain degradation was found and no polymeric material passed through the membranes used for ultrafiltration, the obtained polyelectrolytes **11** and **12** have the same degrees of polycondensation as the precursor PPPs **3** and **6** used for their preparation, *i.e.* PPP polyelectrolytes **11** and **12** became available with values of P_n of 20 - 70 (M_n = 15 000 - 50 000 g·mol^{-1}).

Finally, we would like to emphasize that the PPP polyelectrolytes **9** - **12** are excellently soluble in water, despite their highly hydrophobic skeleton. This is presumably due to the fact that these macromolecules represent cylinders of an average length of l_c = 90 - 300 Å and a diameter of d_c ≈ 10 Å whose hydrocarbon core is covered by the highly polar and thus hydrophilic shell of tetraalkylammonium groups.

Solution properties of PPP polyelectrolytes

Polymer **10** was chosen for all subsequent studies because it is easily available in the gram scale and excellently soluble in water. Moreover, because it bears two charged groups per repeat unit at the ends of the hydrophobic *n*-hexyl side chains, this material was expected to have solution properties almost undisturbed by the hydrophobic moieties of the repeating units. The P_n of polymer **10** discussed here was 40.

a) Viscosimetry

Figure 1 shows the plot of the reduced specific viscosities as function of polymer concentration (Huggins plot) of PPP polyelectrolyte **10** in pure water (■), and in salt-containing solutions having (□) 0.010, (∇) 0.025, and (▲) 0.050 g · L^{-1} of $NEt_4^+ I^-$ as added salt.

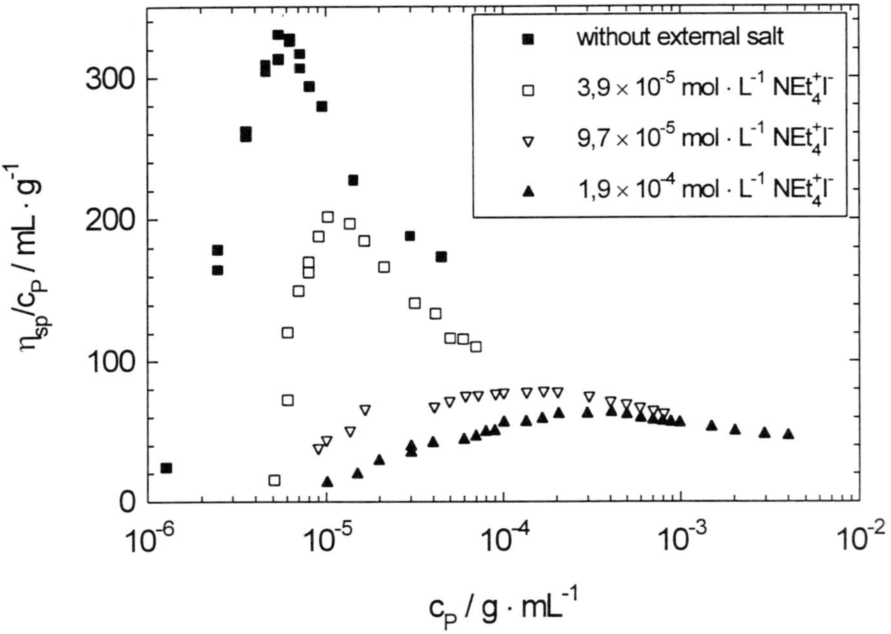

Fig. 1: Huggins plots of PPP polyelectrolyte **10** in pure water (■), and in salt-containing solutions having (□) 0.010, (∇) 0.025, and (▲) 0.050 g · L^{-1} of $NEt_4^+ I^-$.

Without added salt, or at very low concentration of external salt, distinct maxima are found in the Huggins plots of polyelectrolyte **10** which became less and less pronounced with increasing ionic strength. Moreover, the maxima shift from $c_P \approx 5 \times 10^{-6}$ g · mL^{-1} (without added salt) to higher polymer concentrations (such as $c_P \approx 4 \times 10^{-4}$ g · mL^{-1} for 1.9×10^{-4} *M* aqueous $NEt_4^+ I^-$). If the measurements were finally performed at very high concentration of added salt, *i.e.* in 0.01 *M* aqueous $NEt_4^+ I^-$, the data could be fitted by a straight line (not shown). Here, the intrinsic viscosity of the PPP polyelectrolyte **10** under investigation was found to be $[\eta] = 29$ mL · g^{-1}. Due to the high ionic strength, the intermolecular coulombic interactions of the polymer chains are strongly screened out in this latter solution. The intrinsic viscosity of **10** obtained under these conditions is in excellent agreement with the value of [η] measured for its uncharged precursor PPP **3** ([η] = 31 mL · g^{-1}, measured in 1,1,2,2-tetrachloroethane at 35 °C). This result further supports our assumption that changes of the values of P_n during the conversion of the precursor PPPs into the PPP polyelectrolytes are negligible.

The observed hydrodynamic behavior of polymer **10** is in qualitative agreement with recent findings on flexible polyelectrolytes. However, there are marked quantitative differences. For example, the maxima in the Huggins plots of the rodlike PPP polyelectrolytes appear at considerably lower polymer concentrations — approximately one order of magnitude — than the maxima appear in the case of coiled polyelectrolytes of similar chain lengths when measurements are performed in solutions of comparable ionic strength. Moreover, the observed polyelectrolyte effect is much more pronounced in the case of the rodlike PPP polyelectrolytes than in the case of coiled polyelectrolytes of similar chain length and charge density. A detailled analysis of these findings and their quantitative description by means of theoretical models is under way.

b) SAXS investigations

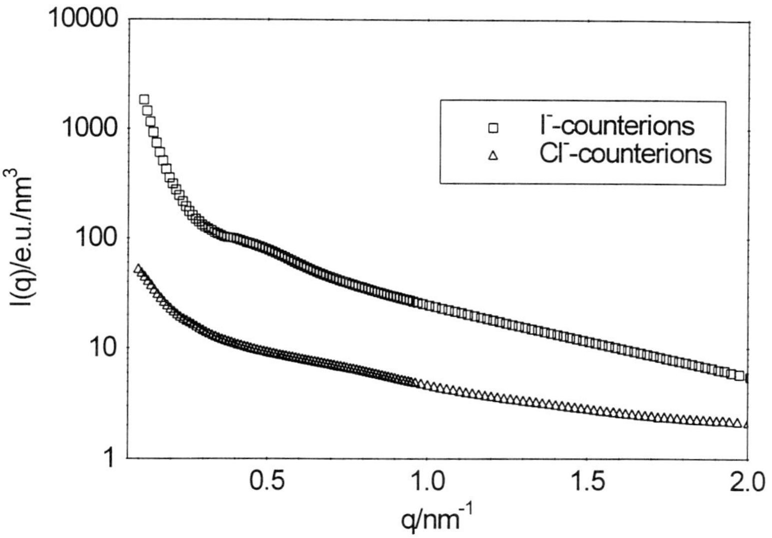

Fig. 2: Absolute SAXS-intensities measured from solutions of polyelectrolyte **10** in aqueous solution at a concentration of 10 g · L^{-1}. Two counterions have been used: Iodine (squares) counterions having a high contrast towards water and chlorine (triangles) counterions which are nearly matched by water.

To assess the correlations of the counterions and the macroions in more detail, the PPP-polyelectrolyte **10** has been studied in solution using small-angle X-rays scattering (SAXS). By changing the counterions from Cl^- to I^- the contrast can be changed quite drastically: While the electron density of Cl^- ions is nearby matched by water, I^- ions exhibit a strong contrast in aqueous solution. Hence, if I^- counterions are used, their correlation with the macroion should be easily visible. In case of Cl^- counterions, on the other hand, the measured scattering intensity is dominated by the macroions.

SAXS measurements in solution were performed as described recently[24)]. Details of measurements and data evaluation will be given elsewhere[25)]. Fig. 2 displays the desmeared SAXS intensities of the polyelectrolyte **10** with I^- counterions (squares) and with Cl^- counterions. The marked difference between both sets of data is immediately obvious: The scattering intensities measured in case of I^- counterions are much higher. This finding points immediately to the strong correlation of the counterions and the macroion: If the I^- counterions were dispersed at random in the solution, their contribution to the measured scattering intensity would be small. If, however, most of the counterions are located in the immediate vicinity of the macoion ("counterion condensation"), strong spatial inhomogeneity of the excess electron density would follow, leading to the observed strong scattering.

Both sets of data can be evaluated to yield quantitative information on the radial distribution of counterions. Hence, these measurements may serve for a quantitative check of current models of rodlike polyelectrolytes as *e.g.* the cell model. A study of this problem by SAXS is under way[25)].

c) Osmotic measurements

The osmotic coefficient of salt-free polyelectrolyte solutions is a classical subject of polymer research[26)]. As in case of the above SAXS data, the resulting data can immediately be compared with the cell model of rodlike polyelectrolytes[27,28)]. In the frame of the cell model, the polyelectrolyte is characterized by the charge parameter[26-28)], ξ, which is defined through the ratio of the Bjerrum length, λ_B, to the contour distance per unit charge, b:

$$\xi = \frac{\lambda_B}{b} \qquad (1)$$

with λ_B being defined through

$$\lambda_B = \frac{e^2}{4\pi\varepsilon_0 \varepsilon k_B T} \qquad (2)$$

where e is unit charge, ε the dielectric constant of the medium and ε_0, k_B and T have their usual meanings. Within the frame of the Manning theory[29,30)] a certain fraction of the counterions is bound to the macroion if ξ exceeds unity. The rest of the counterions remains "free" and forms a Debye-Hückel cloud around the macroion.

For solutions without added salt, the fraction of free counterions may be expressed through the osmotic coefficient ϕ. This quantity is defined by the ratio of the measured osmotic pressure Π and the ideal osmotic pressure Π_{id} calculated for a solution of free counterions

$$\phi = \frac{\Pi}{\Pi_{id}} \qquad (3)$$

In this case, the Manning theory arrives for $\xi > 1$ at a particularly simple result for the osmotic coefficient at infinite dilution, namely[29,30)]

$$\phi = \frac{1}{2\xi} \qquad (4)$$

In the case to be discussed here, *i.e.* solutions of the polyelectrolyte without added salt, the PB-cell model can be solved exactly[31,32)] to yield the osmotic coefficient for finite concentrations of the polyelectrolyte as well:

$$\phi = \frac{1-\beta^2}{2\xi} \qquad (5)$$

Here, β is an integration constant which is related to the concentration of the polyelectrolyte in solution and the radius of the macroion[31,27,28)]. Hence, given these quantities ϕ may be calculated from the cell model[31)] without resort to adjustable parameters. Data of the osmotic coefficient may therefore serve as a sensitive test of the cell model of rigid-rod polyelectrolytes.

Fig. 3 displays the osmotic coefficient measured on a salt-free solution of the PPP polyelectrolyte **10** in water at 40°C. Details of the measurements will be given elsewhere[33)]. The charge parameter has been determined from structural data of the polyelectrolyte **10**. As it turned out, the osmotic coefficient ϕ is not very sensitive towards the diameter a of the macroion which presents the only adjustable parameter for the comparison of theory and experiment. Here we chose a = 0.5 nm which seems to be a reasonable value for the rodlike polyelectrolyte under consideration. Hence, the comparison between the cell model (eq.(5)) and the measured data proceeds virtually parameter-free.

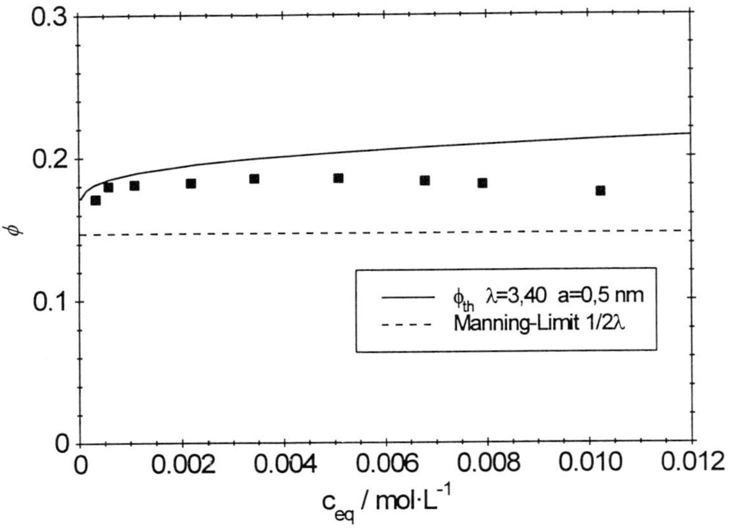

Fig. 3: Osmotic coefficient (see eq.(3)) of salt-free solutions of the polyelectrolyte **10** (squares) together with the theoretical predictions of the cell model[31)] (solid line, eq.(5)) and of the Manning theory[29,30)] (dashed line; eq.(4)). From ref.[33)]

It is evident that the experimental data are located between the prediction of the cell model given by eq.(5) and the Manning limit (eq.(4)). The cell model overestimates ϕ considerably, in particular when going to higher polymer concentrations. Similar findings have been reported from solutions of flexible polyeletrolytes[26)]. Here agreement of theory and experiment was achieved by adjusting the length per unit charge, *i.e.*, by using the charge parameter ξ as an adjustable parameter. In the case of the stiff-chain polyelectrolyte **10** this procedure is no longer possible. The present comparison therefore demonstrates that the theory of the osmotic coefficient is in need of revision. This finding is in agreement with recent theoretical work[34)].

The osmotic coefficient furnished by the Manning theory (dashed line in fig. 3) must be regarded as a limiting value to be found at exceedingly low concentration being not accessible experimentally[34)]. Agreement of the experimental data with ϕ derived from eq.(4) cannot be expected. A more detailed discussion of the osmotic coefficient of the rodlike polyelectrolytes **10** is under way[33)].

Synthesis of stiff-chain ruthenium(II) coordination polyelectrolytes

Soluble, constitutionally well-defined ruthenium(II) coordination polymers are of actual interest because of their electronic, photooptical, magnetic and catalytic properties[35-42]. Moreover, they represent important model systems for the profound analysis of polyelectrolyte behavior in dilute solution. Therefore, we developed efficient synthetic routes leading to conformationally rigid, high-molecular-weight coordination polyelectrolytes **13a-d** and **16**[43-46]. While polymers **13** form random coils and thus are not considered in detail here, polymers **16** are rodlike in shape. Their preparation was possible via the conversion of 4,4''-bis(2,2':6',2''-terpyridine)-2',5'-dihexyl-*p*-terphenyl **14** with an activated ruthenium(III) species such as **15**[47,48] in a solvent mixture of DMA and 1-butanol (Scheme 4). Because of the flexible *n*-hexyl side chains, the dark-red, fibrous rodlike polymers **16** proved to be readily soluble in various organic solvents such as DMA or DMSO, and their constitution could be verified using ^{1}H and ^{13}C NMR spectroscopy: All observed absorptions clearly corresponded to the inner-chain repeating units of **16**, and no absorptions could be found pointing towards defect structures or end-groups. Moreover, by means of model investigations we could prove that these latter absorptions are not only covered by the broad signals of the repeating units but are really absent. Hence, because end-groups absorptions obviously have intensities below the limits of accuracy of the NMR method for polymers **16** prepared from an exact 1:1 stoichiometry of comonomers **14** and **15**, and because formation of cyclic oligomers can be excluded due to the rodlike shape of **16**, we concluded that really high-molecular-weight polymers were produced: Considering the limits of the NMR method, we estimated average degrees of polymerisation of $P_{n(NMR)} \geq 30$ ($M_n \geq 30\ 000$ g · mol^{-1}) for the best products synthesized so far.

Solution properties of ruthenium(II) coordination polyelectrolytes

The hydrodynamic behavior of coordination polyelectrolytes **16** was analyzed both in salt-free and in salt-containing DMA solutions using viscosimetry. In a first series of experiments, polyelectrolytes **16** were studied in 0.02 *M* NH_4PF_6 / DMA. Under these conditions, the intermolecular Coulomb interactions between the charged macromolecules are screened out, and evaluation of the measured data is possible according to the procedures usually applied for uncharged polymers. Fig. 4 displays the Huggins plots of (**A**) a high-molecular-weight sample of **16** ($P_{n,NMR} \geq 30$), of (**B**) a low-molecular-weight sample of **16** ($P_{n,NMR} < 10$), and — for comparison purposes — of (**C**) a high-molecular-weight sample of the coiled ruthenium(II) coordination polymer **13d** ($P_{n,NMR} \geq 30$) bearing adamantylmethylene side groups.

Scheme 4

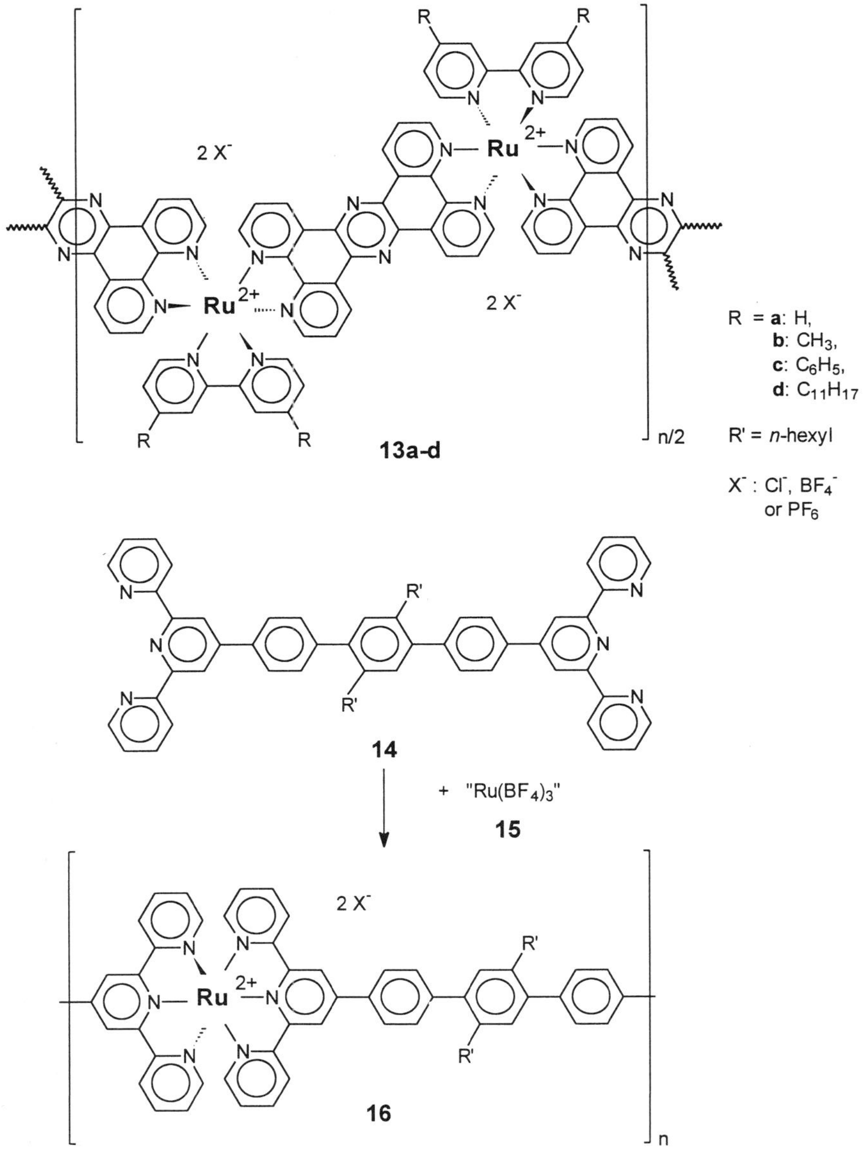

2 X⁻
Ru
2+
R
n/2
13a-d
R = a: H,
b: CH3,
c: C6H5,
d: C11H17
R' = n-hexyl
X⁻ : Cl⁻, BF4⁻ or PF6
R'
14
+ "Ru(BF4)3"
15
16
n

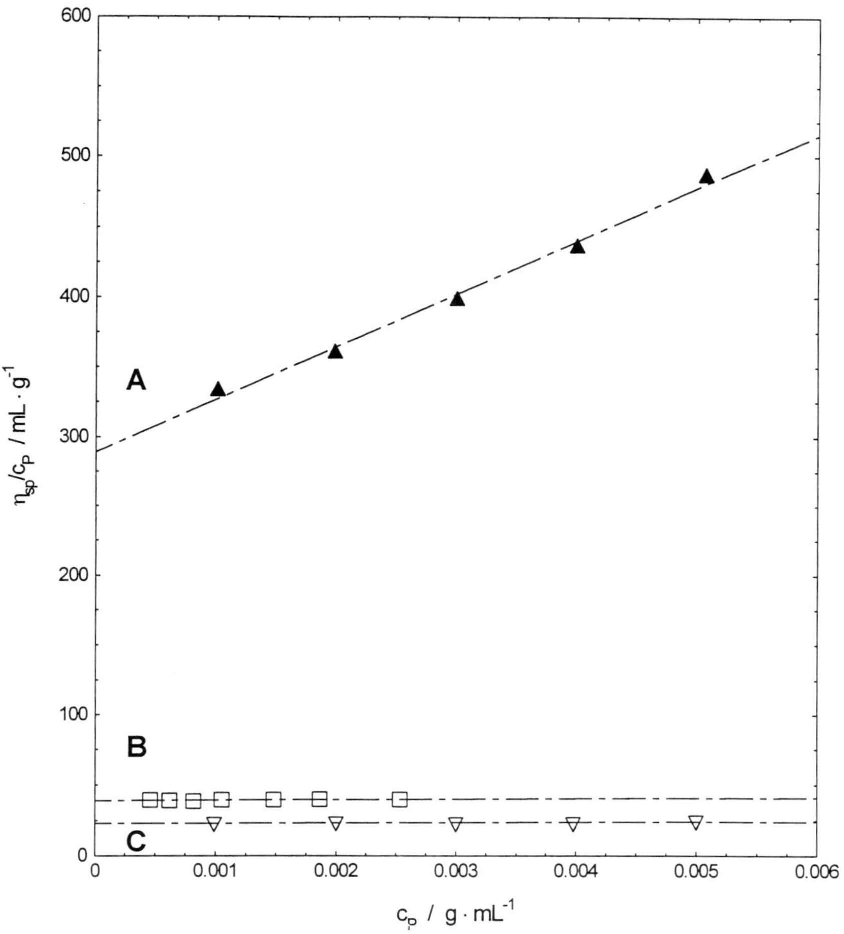

Fig. 4: Huggins plots (**A**) of a high-molecular-weight sample of **16** ($P_{n,NMR} \geq 30$), (**B**) of a low-molecular-weight sample of **16** ($P_{n,NMR} < 10$), and (**C**) of a high-molecular-weight sample of the coiled ruthenium(II) coordination polymer **13d** ($P_{n,NMR} \geq 30$) bearing adamantylmethylene side groups, recorded in 0.02 *M* NH_4PF_6 / DMA solution.

In all three cases, linear extrapolation of the values of η_{sp}/c_P is possible to $c_P = 0$ g · mL^{-1}. An intrinsic viscosity of $[\eta] \approx 300$ mL · g^{-1} is determined for the high-molecular-weight sample of polymer **16**. This value is approx. one order of magnitude higher than the values of $[\eta]$ determined for the rodlike oligomer **16** ($[\eta] \approx 40$ mL · g^{-1}) and for the high-molecular-weight but randomly coiled ruthenium(II) coordination polymer **13d** ($[\eta] \approx 25$ mL · g^{-1}). Moreover, a value of $[\eta] \approx 300$ mL · g^{-1} is not only the highest value ever observed for a ruthenium(II) coordination polymer but is even higher than those found for many other molecular-dispersely soluble, rodlike macromolecules held together by covalent bonds only. Poly(*p*-phenylenes) **17**, for example, have a value of $[\eta] = 207\ mL \cdot g^{-1}$ when the value of P_n is 37 (osmometry)[21,49]. For rodlike polyimides **18**, moreover, a value of $[\eta] = 268\ mL \cdot g^{-1}$ is found for samples having $P_n = 46$ (osmometry) and $P_w = 111$ (static light scattering)[50]. The polyimide **18**, for example, has an average of 46 repeating units per chain and hence an average chain length of $l_n = 96$ nm (the length of the monomeric unit is $l_u = 2.09$ nm)[50]. As the intrinsic viscosity of the sample of **18** under consideration is still somewhat lower than that of **16**, it is reasonable to assume that the average chain length of the novel ruthenium(II) coordination polymers **16** is at least similar. That means if polymer **16** has an average chain length of $l_n \approx 100$ nm as well, its P_n is approx. 43 ($l_u = 2.32$ nm).

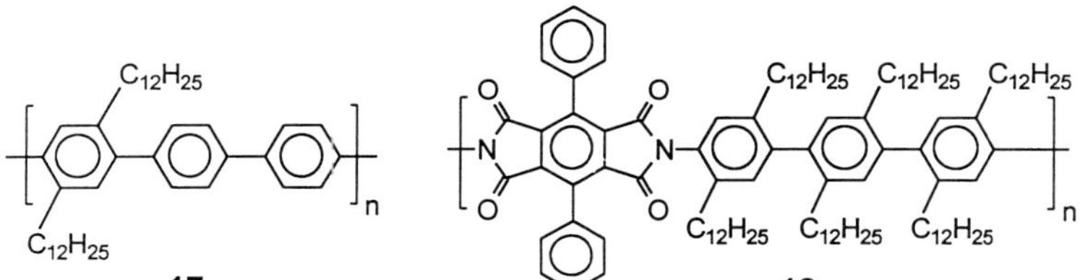

17 **18**

This comparison additionally supports the anticipated high molecular weights of coordination polymers **16**. Even if this procedure does not give the exact values of P_n it nevertheless strongly supports the results from NMR spectroscopy. Moreover, it also supports the rodlike shape of **16**: If it had a substantially flexible or even coiled backbone, neither the high values of the intrinsic viscosity nor the tremendeous influence of the chain length on the intrinsic viscosity would be observed.

In a second series of experiments, the hydrodynamic behavior of coordination polyelectrolytes **16** was studied in pure DMA and thus at low ionic strength. Under these conditions, the intermolecular Coulomb interactions between the charged macromolecules are in operation. Fig. 5 displays the Huggins plot of (**A**) a high-molecular-weight sample of **16** ($P_{n,NMR} \geq 30$), of (**B**) a low-molecular-weight sample of **16** ($P_{n,NMR} < 10$), and of (**C**) a high-molecular-weight sample of the coiled polymer **13a** ($P_{n,NMR} \geq 30$). Under salt-free conditions as well, the high-molecular-weight polymer **16** stands out from the other samples: It displays by far the most pronounced polyelectrolyte effect. A maximum of $(\eta_{sp}/c_P)_{max} \approx 2000\ mL \cdot g^{-1}$ is found for this sample in the Huggins plot at a polymer concentration of $c_P \approx 5 \times 10^{-6}\ g \cdot mL^{-1}$. However, the corresponding maximum is also clearly detectable for the low-molecular-weight rodlike system **16** where a value of $(\eta_{sp}/c_P)_{max} \approx 500\ mL \cdot g^{-1}$ is reached at approx. the same polymer concentration of $c_P \approx 5 \times 10^{-6}\ g \cdot mL^{-1}$. Finally, as expected, the maximum of the reduced specific viscosity of the high-molecular-weight but randomly coiled reference polymer **13a** is again less pronounced $\{(\eta_{sp}/c_P)_{max} \approx 100$ mL · g$^{-1}\}$. It appears at a polymer concentration of $c_P \approx 9\times10^{-5}$ g · mL^{-1} which is more than one order of magnitude higher than the concentration where the maximum is found for polymers **16**.

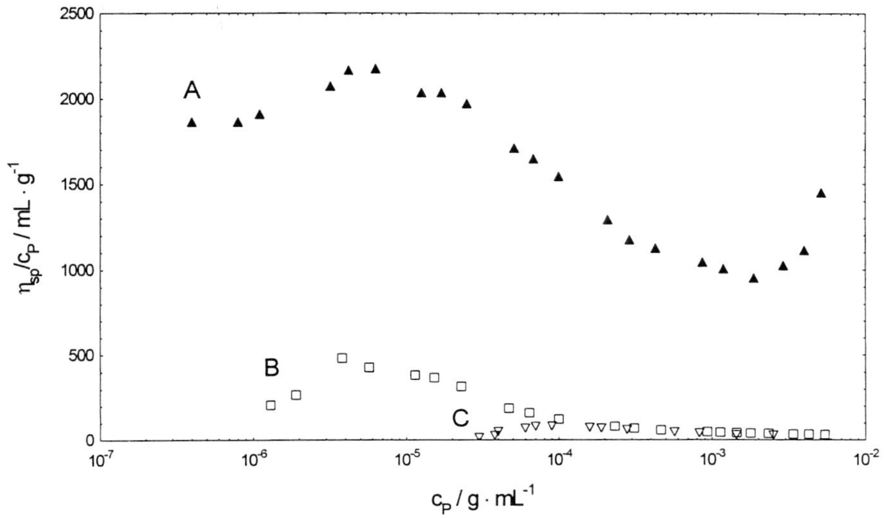

Fig. 5: Huggins plot of (**A**) a high-molecular-weight sample of **16** ($P_{n,NMR} \geq 30$), (**B**) a low-molecular-weight sample of **16** ($P_{n,NMR} < 10$), and (**C**) a high-molecular-weight sample of the coiled polymer **13a** ($P_{n,NMR} \geq 30$), recorded in salt-free DMA solution.

These results highlight the considerable influence of the chain length and in particular of the shape of the macromolecules on their hydrodynamic behavior also in solutions of low ionic strength: Extended chains give rise to much more pronounced polyelectrolyte effects than chains of similar length but coiled conformation. Moreover, while the P_n influences the values of $(\eta_{sp}/c_P)_{max}$ considerably in the case of the rodlike polyelectrolytes **16**, this effect is much less pronounced in the case of the coiled systems **13**.

Finally, comparison of the hydrodynamic behavior of the low-molecular-weight polymer **16** with that of the rodlike PPP polyelectrolyte **10** described above might be instructive. In the latter case, a value of $(\eta_{sp}/c_P)_{max} \approx 350$ mL·g^{-1} is found in salt-free aqueous solution at $c_P \approx 5\times10^{-6}$ g·mL^{-1} in the Huggins plot of a sample **10** having a P_n of 40 (osmometry). This value of P_n corresponds to an average chain length of $l_n \approx 17.6$ nm ($l_u = 0.44$ nm). The low-molecular-weight polymer **16**, on the other hand, which has a $P_{n,NMR} < 10$ and thus an average length of $l_n < 23.2$ nm, shows a maximum in the Huggins plot of $(\eta_{sp}/c_P)_{max} \approx 500$ mL·g^{-1} (Fig. 5, curve **B**). At first glance, these two results fit fairly well. However, prior to a final assessment and more extensive conclusions, the influence of the solvent and of the shear gradient on the observed quantities must be investigated in more detail.

Conclusions

We described the synthesis and the behavior in solution of rigid, rodlike cationic polyelectrolytes having (*i*) poly(*p*-phenylene) backbones and (*ii*) main chains composed of 4,4''-bis(2,2':6',2''-terpyridine)2',5'-dihexyl-*p*-terphenyl moieties and ruthenium(II) centers. All these polymers are shown to have a homogeneous constitution and degrees of polymerization of up to $P_n \approx 70$. Their solution properties could be analyzed using viscosimetry, small-angle X-ray scattering (SAXS) and osmometry. Pronounced polyelectrolyte effects were found in salt-free solutions using viscosimetry. Small-angle X-ray scattering demonstrated directly the strong correlation of the counterions and the macroions. The osmotic coefficients measured in salt-free solution as function of polyelectrolyte concentration were found to be lower than predicted by the cell model.

References

1) S. Förster, M. Schmidt, *Adv. Polym. Sci.* **120**, 51 (1995)
2) K. S. Schmitz, *Macroions in solution and colloid suspension*, VCH Publishers: New York, 1993
3) S. Förster, M. Schmidt, M. Antonietti, *Polymer* **31**, 781 (1990)
4) M. J. MacCallum, C. A. Vincent, *Polymer Electrolyte Reviews*, Elsevier: London, 1987
5) K. Kassapidou, W. Jesse, M. E. Kuil, A. Lapp, S. Egelhaaf, J. R. C. van der Maarel, *Macromolecules* **30**, 2671 (1997)
6) C. C. Lee, S.-G. Chu, G. C. Berry, *J. Polym. Sci.: Polym. Phys. Ed.* **21**, 1573 (1983)
7) P. Metzger Cotts, G. C. Berry, *J. Polym. Sci.: Polym. Phys. Ed.* **21**, 1255 (1983)
8) I. U. Rau, M. Rehahn, *Makromol. Chem. Phys.* **194**, 2225 (1993)
9) I. U. Rau, M. Rehahn, *Polymer* **34**, 2889 (1993)
10) I. U. Rau, M. Rehahn, *Acta Polymerica* **45**, 3 (1994)
11) G. Brodowski, A. Horvath, M. Ballauff, M. Rehahn, *Macromolecules* **29**, 6962 (1996)
12) M. Wittemann, M. Rehahn, *J. Chem. Soc., Chem. Commun.* 623 (1998)
13) R. Rulkens, M. Schulze, G. Wegner, *Macromol. Rapid Commun.* **15**, 669 (1994)
14) S. Vanhee, R. Rulkens, U. Lehmann, C. Rosenauer, M. Schulze, W. Köhler, G. Wegner, *Macromolecules* **29**, 5136 (1996)
15) M. Ballauff, *Mater. Sci. Tech.* **12**, 213 (1993)
16) M. Ballauff, *Angew. Chem.* **101**, 261 (1989)
17) M. Rehahn, A.-D. Schlüter, G. Wegner, W. J. Feast, *Polymer* **30**, 1060 (1989)
18) M. Rehahn, A.-D. Schlüter, G. Wegner, *Makromol. Chem. Phys.* **191**, 1991 (1990)
19) A.-D. Schlüter, G. Wegner, *Acta Polym.* **44**, 59 (1993)

20) G. Brodowski, *PhD thesis*, Karlsruhe, Germany, 1999

21) P. Galda, *PhD thesis*, Karlsruhe, Germany, 1994

22) M. Bohdanecky, *Macromolecules* **16**, 1483 (1983)

23) B. L. Farmer, B. R. Chapman, D. S. Dudis, W. W. Adams, *Polymer* **34**, 1588 (1993)

24) P. Hickl, M. Ballauff, U. Scherf, K. Müllen, P. Lindner, *Macromolecules*, **30**, 273 (1997)

25) B. Guilleaume, M. Ballauff, in preparation

26) M. Mandel, *Polyelectrolytes*. In *Encyclopedia of Polymer Science and Engineering*, 2nd edition, F. H. Mark, N. M. Bikales, C. G. Overberger, G. Menges, Eds., Wiley, New York, 1988, Vol. 11, p. 739

27) A. Katchalsky, *Pure Appl. Chem.* **26**, 327 (1971) and further references given there

28) M. Mandel, *J. Phys. Chem.* **96**, 3934 (1992)

29) G. S. Manning, *J. Chem. Phys.* **51**, 924, 934, 3249 (1969)

30) G. S. Manning, *Ann Rev. Phys. Chem.* **23**, 117 (1972)

31) R. M. Fuoss, A. Katchalsky, S. Lifson, *Proc. Natl. Acad. Sci. U.S.A.* **37**, 579 (1951)

32) T. Alfrey, P. W. Berg, H. Morawetz, *J. Polym. Sci.* **7**, 543 (1951)

33) J.Blaul, M. Ballauff, in preparation

34) For the possible reasons of this failure of the cell model, see P. S. Kuhn, Y. Levin, M. C. Barbosa, *Macromolecules* **31**, 8347 (1998)

35) M. Rehahn, *Acta Polymerica* **49**, 201 (1998)

36) F. Ciardelli, E. Tsuchida, D. Wöhrle, *Macromolecule-Metal-Complexes*, Springer, Berlin, 1996

37) V. Balzani, A. Juris, M. Venturi, S. Campagna, S. Serroni, *Chem. Rev.* **96**, 759 (1996)

38) A. Harriman, R. Ziessel, *J. Chem. Soc., Chem. Commun.* 1707 (1996)

39) J.-M. Lehn, *Supramolecular Chemistry*, VCH, Weinheim, 1995

40) E. C. Constable, A. M. W. Chargill Thompson, *J. Chem. Soc., Dalton Trans.* 1615 (1995)

41) E. C. Constable, A. M. W. Cargill Thompson, P. Harveson, L. Macko, M. Zehnder, *Chem. Eur. J.* **1**, 360 (1995)

42) J.-P. Sauvage, J.-P. Collin, J.-C. Chambron, S. Guillerez, C. Coudret, V. Balzani, F. Barigelletti, L. DeCola, L. Flamigni, *Chem. Rev.* **94**, 993 (1994)

43) R. Knapp, A. Schott, M. Rehahn, *Macromolecules* **29**, 478 (1996)

44) S. Kelch, M. Rehahn, *Macromolecules* **30**, 6185 (1997)

45) S. Kelch, M. Rehahn, *Macromolecules* **31**, 4102 (1998)

46) S. Kelch, M. Rehahn, manuscript submitted

47) M. Beley, S. Chodorowski, J.-P. Collin, J.-P. Sauvage, L. Flamigni, F. Barigelletti, *Inorg. Chem.* **33**, 2543 (1994)

48) W. Spahni, G. Calcaferri, *Helv. Chim. Acta* **67**, 450 (1984)

49) M. Rehahn, A.-D. Schüter, G. Wegner, *Makromol. Chem.* **191**, 1991 (1990)

50) L. Schmitz, M. Ballauff, *Polymer* **36**, 879 (1995)

Synthesis of Nonlinear Optical Polymers with Large Photoconductive Sensitivity and Transparency

Seong Yong Pyun, Hyunsik Moon, Jin Kyun Lee, Nakjoong Kim,[+]
Soo Young Park*

Department of Fiber and Polymer Science, Seoul National University, San 56-1, Shillim-dong, Kwanak-gu, Seoul 151-742, Korea
[+] Electronic Materials and Devices Research Center, Korea Institute of Science and Technology, P.O. Box 131, Cheongryang, Seoul 136-650, Korea
*To whom correspondence should be addressed.

SUMMARY: Novel nonlinear optical (NLO) chromophore, 2-{3-[2-(4-methylsulfonylphenyl)vinyl]carbazol-9-yl}ethanol was synthesized and subsequently reacted with methacryloyl chloride to give a photoconducting NLO monomer (**M1**). 2-Methylacrylic acid 2-[3-(diphenylhydrazonomethyl)carbazol-9-yl]ethyl ester (**M2**) was also synthesized as a comonomer to enhance the carrier mobility of the NLO polymer. Photoconducting NLO polymers, **P1** and **P2** were obtained by the copolymerization of **M1** with methyl methacrylate and **M2**, respectively. These polymers were well soluble in organic solvents and showed glass transition at 177 ℃ and 196 ℃, respectively. Polymer films of **P1** and **P2** were optically clear, and were transparent at wavelengths longer than 420 nm. The electro-optic coefficient (r_{33}) of poled **P1** films was measured to be ~5 pm/V at 632.8 nm. The photoconductive sensitivities of **P1** and **P2** were 6.2×10^{-14} S·cm^{-1}/mW·cm^{-2} and 5.6×10^{-11} S·cm^{-1}/mW·cm^{-2}.

Introduction

Photorefractive materials are potentially useful in the high-density optical data storage, optical image processing, dynamic holography and optical computing.[1,2] Photoconducting nonlinear optical (NLO) polymers are the most promising candidates for these photorefractive applications.[3,4] Photoconductivity and second-order NLO activity, necessary to implement the photorefraction, are normally provided by the respective chromophores, which are chemically attached to polymer backbone or simply dispersed as a guest. Fully functionalized polymers with chemically attached chromophores are very stable and reliable for the photorefractive applications.[5,6] However, they suffer from the difficulties in the synthesis, the control of copolymer composition, and consequently the optimization of photorefractivity. Therefore, we are rather interested in the use of a single monolithic chromophore, namely photoconducting NLO chromophore, for the preparation of corresponding side-chain polymer.

 CCC 1022-1360/99/$ 17.50+.50/0

We have demonstrated that the heteroaromatic compounds like carbazole,[7,8] indole,[9] or indoline[10] with 2-(4-nitrophenyl)vinyl substituent at 3 or 5-position are examples of monolithic chromophore. It was shown that the polymers containing these monolithic chromophores as side chain exhibited excellent NLO activity and moderate photoconductivity.[7-10] Two main problems were realized when we try to utilize these polymers as photorefractive media. First, the absorption maxima (λ_{max}'s) of these chromophores were located at the wavelength longer that 420 nm with their apparent cut-off usually extended into 550 nm. Such an insufficient transparency in visible region is a real problem, since even a small tailing absorption at the wavelength of laser source (for example, 532 nm or 633 nm) is sufficient to prevent the polymer from getting photorefractive net gain. Second, the photoconductive sensitivity of monolithic chromophore was not high enough to assure fast response.
In this work, we report the appropriate molecular design and synthesis to solve these problems. To afford better transparency in the visible region, we synthesized monolithic chromophore, 2-{3-[2-(4-methylsulfonylphenyl)vinyl]carbazol-9-yl}ethanol, with apparent cut-off at 420 nm. To improve the photoconductive sensitivity, we synthesized 2-methylacrylic acid 2-[3-(diphenylhydrazonomethyl)carbazol-9-yl]ethyl ester as a comonomer of photoconductive NLO polymer.

Experimental

Instruments

^{1}H-NMR spectra were recorded with the use of Bruker AMX 300 spectrometer. IR spectra were measured with KBr pellet or KBr windows on Midac FT-IR spectrophotometer. Mass spectrometer data were obtained from JMS AX505WA using electron impact (EI) mode. Shimadzu UV 2101PC double beam spectrophotometer was used for the UV-Vis absorption spectra. Gel permeation chromatography (GPC) was performed at the flow rate of 1.0 mL/min in THF at 30 ℃ with Waters HPLC component system equipped with five Ultra-μ-styragel® columns (2×10^5, 10^5, 10^4, 10^3, 500 Å), which was calibrated with polystyrene standards. Differential scanning calorimetry (DSC) and thermogravimetric analysis (TGA) were carried out under nitrogen atmosphere at the heating rate of 10 ℃/min on Perkin Elmer DSC7 and Du Pont 2950, respectively.

Synthesis of Monomers

Synthetic pathway to the photoconducting nonlinear optical monomer, 2-methylacrylic acid 2-

{3-[2-(4-methylsulfonylphenyl)vinyl]carbazol-9-yl}ethyl ester (**M1**), and the comonomer for enhanced carrier mobility, 2-methylacrylic acid 2-[3-(diphenylhydrazonomethyl)-carbazol-9-yl]ethyl ester (**M2**) is outlined in Schemes 1.

2-(Carbazole-9-yl)ethanol (**1**), acetic acid 2-(3-formylcarbazol-9-yl)ethyl ester (**2**) and 9-(2-hydroxyethyl)-9H-carbazole-3-carbaldehyde (**3**) were prepared as reported earlier.[7]

4-(Methylthio)benzyl chloride (4). 4-(Methylthio)benzyl alcohol (2.00 g, 13.0 mmol) was dissolved in 10 mL of dry toluene and thionyl chloride (1.42 mL, 19.5 mmol) was added dropwise at room temperature. The reaction mixture was stirred at 80 ℃ for 2 h, and then cooled to room temperature. Toluene and excess thionyl chloride were removed under reduced pressure. The crude product was distilled in vacuum to yield 1.48 g of faint yellow liquid (66%). ^{1}H-NMR ($CDCl_3$) 2.37 (s, 3H, SCH_3), 4.44 (s, 2H, $ClCH_2$), 7.12 (d, 2H, aromatic protons), 7.18 (d, 2H, aromatic protons).

Diethyl 4-(methylthio)benzyl phosphonate (5). To the 10 mL of triethylphosphite heated to reflux, **4** (3.00 g, 17.4 mmol) was added dropwise and refluxed for additional 2 h. After cooling, the excess triethylphosphite was removed under reduced pressure. The product was purified by silica gel column chromatography (ethylacetate/n-hexane = 2/1) to yield 4.31 g of water clear liquid (90%). ^{1}H-NMR ($CDCl_3$) 1.23 (t, 6H, CH_2CH_3) 2.47 (s, 3H, SCH_3), 3.07 (d, 2H, PCH_2), 3.97 (q, 4H, CH_2CH_3), 7.18 (t, 4H, aromatic protons).

Diethyl 4-(methylsulfonyl)benzyl phosphonate (6). To a magnetically stirred solution of **5** (10.0 g, 36.5 mmol) in 30 mL of glacial acetic acid was added hydrogenperoxide (30% in water, 10.4 mL, 91.3 mmol) at room temperature. After refluxing for 2 h, the solvent was removed under reduced pressure. The product was purified by silica gel column chromatography (methanol/ethylacetate = 1/6) to yield 6.56 g of white solid (59%). ^{1}H-NMR ($CDCl_3$) 1.27 (t, 6H, CH_2CH_3), 3.06 (s, 3H, SO_2CH_3), 3.21 (d, 2H, PCH_2), 4.02 (q, 4H, CH_2CH_3), 7.50 (d, 2H, aromatic protons), 7.88 (d, 2H, aromatic protons).

2-{3-[2-(4-Methylsulfonylphenyl)vinyl]carbazole-9-yl}ethanol (7). A magnetically stirred solution of **6** (3.37 g, 11.0 mmol) in 40 mL of tetrahydrofuran (THF) was treated with potassium *t*-butoxide (1.85 g, 16.5 mmol) at room temperature. A solution of **2** (3.00 g, 11.0 mmol) in 10 mL of THF was added and then the reaction mixture was heated at reflux for 30 min. After cooling, potassium hydroxide (2.46 g, 44.0 mmol) and 40 mL of ethanol were added and the solution was heated again at reflux for 30 min. The solution was cooled and the solvent was removed under reduced pressure. The residue was poured onto ice-cold water and neutralized with 1 N HCl solution. The precipitate was collected by suction-filtration and the resulting yellow solid was recrystallized from chloroform/ethanol (1/2) to yield 2.19 g of greenish yellow crystal (51%). mp 164 ~ 166 ℃; ^{1}H-NMR ($CDCl_3$) 1.64 (br, 1H, alcohol proton), 3.07 (s, 3H, SO_2CH_3), 4.10 (t, 2H, NCH_2), 4.51 (t, 2H, OCH_2), 7.14 (d, 1H, $J = 16.3$, stilbene proton), 7.49 (d, 1H, stilbene proton); IR (KBr, cm^{-1}) 3515 (-OH), 1587 (-CH= CH),1290, 1151 (-SO_2-).

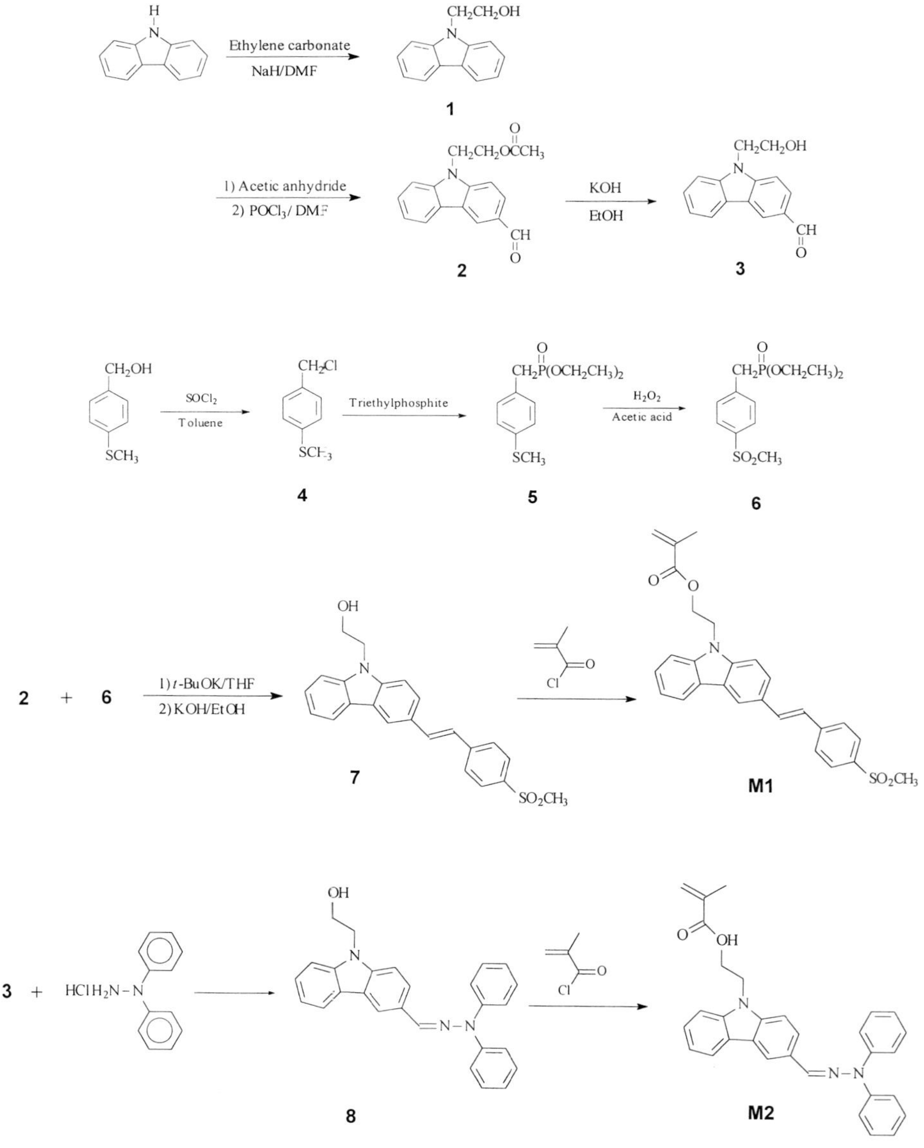

Scheme 1. Synthesis of monomers (**M1** and **M2**).

1-[3-(Diphenylhydrazonomethyl)carbazol-9-yl]ethanol (8). To a magnetically stirred solution of 1,1-diphenylhydrazine hydrochloride (2.77 g, 12.5 mmol) in 100 mL of ethanol was added sodium acetate trihydrate (2.56 g, 18.8 mmol) at room temperature. And then a solution of **3** (2.50 g, 10.4 mmol) in 100 mL ethanol was added dropwise for 30 min. After stirring for 3 h, ethanol was removed under reduced pressure and the residue was poured onto water. The product was extracted with chloroform, washed with plenty of brine and distilled water, and dried over $MgSO_4$. The solvent, chloroform was removed under reduced pressure. The dark brown crude product was purified by silica gel column chromatography (ethyl acetate/n-hexane = 1/1) to yield 4.02 g of dark yellow solid (94%). mp 81 ~ 83 ℃; ^{1}H-NMR ($CDCl_3$) 3.94 (t, 2H, CH_2N), 4.36 (t, 2H, CH_2O), 7.09 ~ 7.16 (m, 7H, aromatic protons), 7.26 ~ 7.36 (m, 8H, aromatic protons), 7.75 (d, 1H, aromatic proton), 7.96 (d, 1H, aromatic proton), 8.10 (s, 1H, aromatic proton); IR (KBr, cm^{-1}) 3388 (-OH), 1597 (C=N).

2-Methylacrylic acid 2-{3-[2-(4-methylsulfonylphenyl)vinyl]carbazol-9-yl}ethyl ester (M1). To a magnetically stirred solution of **7** (2.00 g, 5.12 mmol) in 100 mL dry methylene chloride was added triethylamine (1.73 mL, 15.4 mmol) and methacryloyl chloride (1.00 mL, 10.2 mmol) at 0 ℃. The reaction mixture was allowed to warm up to room temperature and was stirred for 30 min. The solution was washed with saturated sodium bicarbonate solution, and dried over $MgSO_4$. The solvent was removed under reduced pressure. The yellow residue was purified by silica gel column chromatography (ethyl acetate/chloroform = 1/15) to give 1.67 g of yellow solid (71%). mp 181 ~ 183 ℃; ^{1}H-NMR ($CDCl_3$) 1.74 (s, 3H, $C=CCH_3$), 3.01 (s, 3H, SO_2CH_3), 4.48 (t, 2H, NCH_2), 4.55 (t, 2H, OCH_2), 5.42 (s, 1H, vinyl proton), 5.87 (s, 1H, vinyl proton), 7.08 (d, 1H, J = 16.3, stilbene proton), 7.36 (d, 1H, stilbene proton); IR (KBr, cm^{-1}) 1716 (ester C=O), 1630 (vinyl group), 1587 (-CH=CH-), 1290, 1151 ($-SO_2-$); m/z (EI, high resolution) calc'd for $C_{27}H_{25}O_4N_1S_1$, 459.15, found 459.15.

2-Methylacrylic acid 2-[3-(diphenylhydrazonomethyl)carbazol-9-yl]ethyl ester (M2). This monomer was prepared from **8** (5.00 g, 12.3 mmol), triethylamine (5.20 mL, 37.0 mmol), and methacryloyl chloride (2.40 mL, 24.6 mmol) in 50 mL dry of methylene chloride by the same procedure as described for **M1** (4.70 g of yellow solid, 81%). mp 121 ~ 123 ℃; ^{1}H-NMR ($CDCl_3$) 1.80 (s, 3H, $C=CCH_3$), 4.52 (t, 2H, CH_2N), 4.57 (t, 2H, CH_2O), 5.47 (s, 1H, vinyl proton), 5.92 (s, 1H, vinyl proton), 7.17 ~ 7.27 (m, 7H, aromatic protons), 7.38 ~ 7.47 (m, 8H, aromatic protons), 7.86 (d, 1H, aromatic proton), 8.07 (d, 1H, aromatic proton), 8.20 (s, 1H, aromatic proton); IR (KBr, cm^{-1}) 1712 (ester C=O), 1631 (vinyl group), 1597 (-C=N-); m/z (EI, high resolution) calc'd for $C_{31}H_{27}N_3O_2$, 473.21, found 473.21.

Polymerization

Photoconducting NLO polymers, **P1** and **P2** were obtained by the copolymerization of **M1** with methyl methacrylate and **M2**, respectively. A mixture of **M1** (1.2 g, 2.6 mmol), methyl methacrylate (MMA, 0.85 g, 8.0 mmol.), 2,2-azobisisobutyronitrile (AIBN, 12 mg, 0.7 mol%) and 15 mL of dried N-methyl pyrrolidinone (NMP) was degassed by standard freeze-pump-thaw method. After sealing the ampoule, the reaction mixture was heated at 65 ℃ for 24 h. After cooling, the resulting solution was cooled to room temperature, diluted to twice original volume with chloroform and then poured into methanol to precipitate the polymer **P1**, which was purified by repeated reprecipitation from chloroform to methanol. 1.7 g of **P1** was obtained (83%).

P2 was similarly copolymerized. The molar ratio of **M1** (1.30 g, 2.80 mmol) to **M2** (1.34 g, 2.80 mmol) was 1:1. NMP (15 mL) and AIBN (0.7 mol%) were used as solvent and initiator respectively (2.0 g, 76%).

Fabrication of Polymer Film

The prepared polymers (**P1** and **P2**) were dissolved in tetrachloroethane (10 wt%) and the solution was filtered with a Teflon membrane filter (Millipore, 0.22 ㎛). Thin polymer films were spin-coated at 1500 rpm onto indium-tin-oxide (ITO) coated glass. Residual solvent was removed by heating the films in vacuum oven for several days. Film thickness was measured to be about 1.2 ㎛ with TENCOR P-10 profilometer.

Electro-Optic Coefficient

For the measurement of electro-optic coefficient (r_{33}), contact poling was carried out. A gold electrode was sputter-coated in 10^{-6} Torr vacuo on the polymer film backed with ITO glass. To align the chromophores, the sample was heated and then appropriate poling voltage was applied to the gold electrode. After 15 min, the film was cooled down to about 30 ℃ and the poling voltage was removed. The electro-optic coefficient of the poled polymer was measured at the wavelength of 632.8 nm (He-Ne laser) using a simple reflection method.[11]

Photoconductivity

The dark- and the photo- conductivities of polymer films were evaluated by measuring a current through the polymer film sandwiched between gold and ITO using Keithley 617 electrometer. ORIEL 300 W Xe lamp was used as a light source. Positive voltage was applied to ITO electrode and the light was shed through ITO electrode. Photoconductive sensitivity was calculated as the ratio of photoconductivity to the light intensity.

Results & Discussions

Synthesis of Monomers

Photoconducting NLO chromophore, 2-{3-[2-(4-methylsulfonylphenyl)vinyl]carbazol-9-yl}-ethanol (**7**) was obtain by Wadsworth and Emmons reaction between acetic acid 2-(3-formyl-carbazol-9-yl)ethyl ester (**2**) and diethyl 4-(methylsulfonyl)benzyl phosphonate (**6**) in 51% yield. Compound **2** was synthesized from carbazole in three steps, i.e. hydroxyethylation, protection of hydroxy group, and Vilsmeier formylation in 57% overall yield. Compound **6**, was prepared according to the reported method,[12] which consists of three sequential steps starting from 4-(methylthio)benzyl alcohol, namely, chlorination, phosphonation and oxidation in 35% overall yield. Photoconducting NLO monomer (**M1**) was prepared in 71% yield by the reaction of **7** with methacryloyl chloride.

2-Methylacrylic acid 2-[3-(diphenylhydrazonomethyl)carbazol-9-yl]ethyl ester (**M2**), was synthesized from **3** and 1,1-diphenylhydrazine hydrochloride, and subsequent reaction with methacryloyl chloride in 76% overall yield.

Physical Properties of Polymers

Photoconducting NLO polymers, **P1** and **P2**, were prepared by free radical polymerization. Tab. 1 summarizes the molecular weights and glass transition temperatures of them. Polymers are highly soluble in common solvents such as chloroform, dichloroethane, DMF and NMP. However, they were partially soluble in THF. Due to this good solubility and moderate molecular weight as listed in Tab. 1, optical quality thin films could be easily obtained by spin coating. Thermal properties of the polymers were examined by DSC and TGA measurements. From the DSC thermograms, glass transition temperatures (T_g's) of **P1** and **P2** were found to be 177 ℃ and 196 ℃, respectively. Compared to the T_g of PMMA (120 ℃), those of **P1** and **P2** are quite high presumably due to the bulkiness of the carbazole-based chromophore which severely restrict main-chain and side-group motions in these polymers. TGA thermograms showed that the polymers were relatively stable up to 220 ℃ (the temperature of 2% weight loss, under nitrogen).

As expected from the weaker electron-withdrawing ability of sulfone group than that of nitro group, π - π^* absorption bands of **P1** and **P2** in UV-Vis spectrum (Fig. 1) were significantly blue-shifted relative to the nitro-containing photoconducting NLO polymers reported earlier by us.[7-10] The amount of blue-shift was larger than 80 nm, which provided the polymers (**P1** and **P2**) with a broad transparency window in the visible region.

Tab. 1. Properties of polymers

Polymer	m_{feed} [a)]	conversion (%)	M_n [b)]	M_w / M_n	T_g (℃) [c)]
P1	1 : 3 (**M1**: MMA)	83	42700	3.03	177
P2	1 : 1 (**M1** : **M2**)	76	28400	4.29	196

[a)] Mole ratio of monomers in the feed.
[b)] Measured by GPC in THF using polystyrene standards.
[c)] Determined by DSC at a heating rate of 10 ℃/min under N_2.

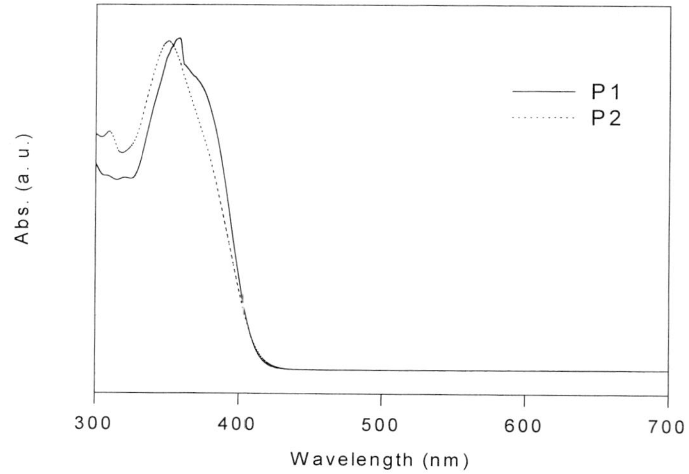

Fig. 1: UV-VIS spectra of **P1** and **P2**

NLO and Photoconducting Properties

First hyperpolarizability (β) of chromophores was calculated by MOPAC 6 using PM3 Hamiltonian. The β value of photoconducting NLO chromophore (**7**) was 18.0×10^{-30} esu, which was about half of that of well-known 4′-(dimethylamino)-4-nitrostilbene (**DANS**). To investigate the dependence of electro-optic coefficient (r_{33}) on the poling field, we poled polymer film of **P1** at 120 ℃ under the electric field of 0.2, 0.4 and 0.6 MV/cm, respectively. As shown in Figure 2, r_{33} increased linearly with the poling field and 5 pm/V was achieved at the electric field of 0.6 MV/cm. The r_{33} value is expected to increase further by the optimization of poling temperature and field.

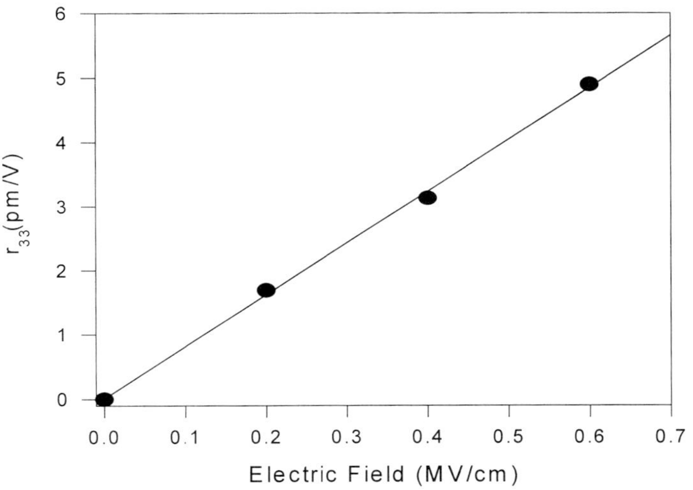

Fig. 2: Dependence of electro-optic coefficient (r_{33}) of **P1** on poling field.

When 3.38 mW/cm^2 light of Xe lamp was irradiated on the polymer films of **P1** and **P2**, orders of magnitude increase in the current density was observed. Photoconductivity values of polymer films were calculated from these photocurrent densities as a function of electric field as shown in Figure 3. Estimated photoconductivity values were 2.1×10^{-13} $S\cdot cm^{-1}$ and 1.9×10^{-10} $S\cdot cm^{-1}$ at the electric field of about 0.6 MV/cm for **P1** and **P2**, respectively. It should be noted that **P2** has 1000 times higher photoconductivity than **P1,** though their absorption spectra are virtually same (Fig. 1), which strongly suggests that **P2** possesses more efficient charge transporting property than **P1**. This higher photoconductivity of **P2** is, no doubt, due to **M2**, which contains carbazole-hydrazone moiety with enhanced carrier transporting property. The calculated photoconductive sensitivities are 6.2×10^{-14} $S\cdot cm^{-1}/mW\cdot cm^{-2}$ for **P1** and 5.6×10^{-11} $S\cdot cm^{-1}/mW\cdot cm^{-2}$ for **P2**, under the 3.38 mW/cm^2 irradiation from Xe lamp.

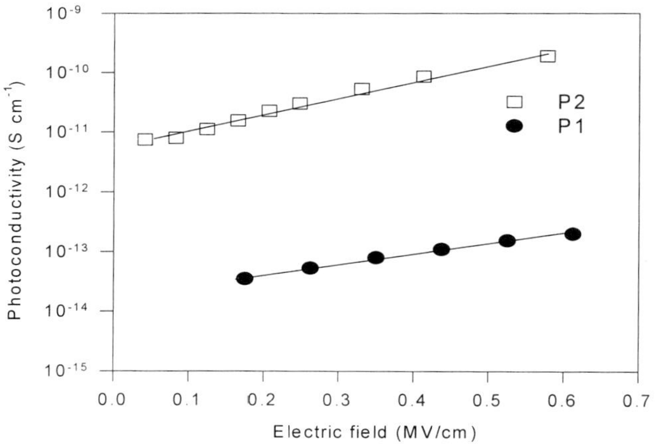

Fig. 3: Photoconductivities of **P1** and **P2**.

Conclusion

We synthesized photoconducting NLO side-chain polymers with large photoconductive sensitivity and broad transparency in visible region. 2-{3-[2-(4-Methylsulfonylphenyl)vinyl]-carbazole derivative was synthesized as photoconducting NLO chromophore with absorption maximum at 350 nm and apparent cut-off at 420 nm. 2-Methylacrylic acid 2-[3-(diphenyl-hydrazonomethyl)carbazol-9-yl]ethyl ester was also utilized as a comonomer to increase photoconductive sensitivity of photoconducting NLO polymer by thousand times.

Acknowledgement

This work was supported by Korea ministry of education research fund for advanced materials in 1996 (RIAMI-NE27-96).

References

1. P. C. Yeh, "*Introduction to Photorefractive Nonlinear Optics*", John Wiley & Sons, New York 1993.
2. D. D. Nolte, "*Photorefractive Effects & Materials*", Kluwer Academic Publishers, Boston 1995.
3. L. Yu, W. K. Chan, Z. Peng, W. Li, A. R. Gharavi, in "*Handbook of Organic Conductive*

Molecules and Polymers", edited by H. S. Nalwa, John Wiley & Sons, Chichester 1997, **4**, Chap. 5, pp 233-260.
4. Y. Zhang, T. Wada, H. Sasabe, *J. Mater. Chem.* **8**, 809 (1998).
5. Z. Peng, A. R. Gharavi, L. Yu, *J. Am. Chem. Soc.* **119**, 4622 (1997).
6. M. S. Bratcher, M. S. DeClue, A. Grunnet-Jepsen, D. Wright, B. R. Smith, W. E. Moerner, J. S. Siegel, *J. Am. Chem. Soc.* **120**, 9680 (1998).
7. D. W. Kim, S. I. Hong, S. Y. Park, N. Kim, *Bull. Kor. Chem. Soc.* **18**, 198 (1997).
8. D. W. Kim, H. Moon, S. Y. Park, S. I. Hong, *Reactive & Functional Polymers*, in press.
9. H. Moon, S. Y. Park, unpublished result.
10. H. Moon, N. Kim, S. Y. Park, *Nonlinear Optics*, in press.
11. C. C. Teng , H.-T. Man, *Appl. Phys. Lett.* **56**, 30 (1990).
12. A. Ulman, C. S. Willams, W. Kohler, D. R. Robello, D. J. Williams, L. Handley, *J. Am. Chem. Soc.* **112**, 7083 (1990).

Synthesis and Properties of New Aromatic Polyimides – Polymeric Materials for Light-emitting and Liquid-crystal Displays

Moonhor Ree,* Sang Il Kim, Seung Moon Pyo, Tae Joo Shin, Hong Ki Park, Jin Chul Jung[1]

Department of Chemistry, Polymer Research Institute, and [1]Department of Materials Science & Engineering, Pohang University of Science & Technology, San 31, Hyoja-dong, Pohang 790-784 Korea

SUMMARY: New aromatic diamines containing π-conjugated heterocyclic rings were synthesized, which emit blue light in the photoluminescence. Using these new monomers, soluble poly(amic acid)s were prepared by the polycondensations with pyromellitamic dianhydride, and thermally converted to the polyimides in films which emit intense blue-light. These polyimides are considered as potential candidate materials for fabricating optoelectronic devices which emit blue light. In addition, new poly(*m*-phenylene 4,4'-oxydiphthalimide)s containing various side chains were synthesized. For these polyimides, thermal properties were investigated with considering the chain flexibility of polymer backbone as well as the spacer and biphenyl mesogen end group in the side chains. In particular, these polyimides exhibited excellent performance in the rubbing process and the controlling of both the alignment and the pretilt of liquid-crystal (LC) molecules in the LC cell. This might be attributed mainly to a strong interaction between the biphenyl mesogen end group in the side chains and the mesogen unit of LC molecules.

Introduction

With increasing the usage of portable computers and telecommunication devices, there are highly demanded flat panel displays which have light weights and are operative with low power. Light-emitting display (LED) is counted as a good example of flat panel displays. A high performance light-emitting material is critical to fabricate LEDs. In particular, light-emitting polymeric materials have recently gained great attention because of their mass productionability and easy processability. Since a diode fabricated with poly(*p*-phenylenevinylene) (PPV) as the emitting layer was first reported to emit green light in 1990[1)], a number of polymeric materials suitable to LEDs have been developed. However, the most efficient polymeric systems are known to be based on PPV films, which emit basically green light rather than blue light (namely, one of the essential three light elements). There is still a big challenge in developing polymer systems to exhibit highly efficient blue-light emission, which can not be achieved easily from inorganic materials.

Another representative flat panel display is the liquid-crystal display (LCD). For full-color LCDs, polyimide films rubbed with a cloth are widely used to control the alignment of

 CCC 1022-1360/99/$ 17.50+.50/0

LCs in contact with the polymer film. In the case, LCs are known to be oriented in the rubbing direction. This phenomenon is directly related to the performance of LCDs, so that numerous research groups have involved to investigate its mechanism. However, the mechanism of aligning LCs at the rubbed polymer surface, especially at the rubbed surface of polymers with side chains, has been not understood fully, so that it is still one of the main subjects to be clarified in the field of LCDs.

In this study, 5,5'-bis(4-aminophenyl)-2,2'-bifuryl and 2,2'-bis(furyl)benzidine were synthesized, which are the new blue-light-emitting diamine monomers with well-defined conjugation lengths. Soluble poly(amic acid)s were synthesized by the polycondensations of these diamines with pyromellitic dianhydride. The soluble precursors then were thermally converted to the corresponding polyimides. The uv-visible absorption and fluorescence characteristics were investigated. And, the thermal stabilities were examined.

In addition, a series of new polyimide precursors were synthesized from 4,4'-oxydiphthalic anhydride and *m*-phenylene diamines containing various side chains. These side chains consist of a flexible alkyl chain spacer linked to various positions of the biphenyl mesogen end group which possibly plays together with LCs in the alignment. Thermal imidization behaviors of the precursor polymers were examined by thermogravimetry. The thermal properties were investigated by thermogravimetry, thermomechanical analysis and differential scanning calorimetry. And, the polyimide films were treated by rubbing with a cloth and used as alignment layers for constructing LC cells. In the LC cell, the pretilt characteristics of LCs were studied by crystal rotation technique as a function of rubbing density. All the results were understood with considering the role of mesogen end group and the side chain conformations and lengths in addition to the polymer backbone structure.

Experimental

Synthesis of monomers. 5,5'-Bis(4-aminophenyl)-2,2'-bifuryl (PFDA) was synthesized from 4-nitroaniline as a starting material, according to the synthetic scheme shown in Fig. 1. In the synthesis, *p*-nitrobenzenediazonium tetrafluoroborate **(1)** was obtained with a yield of 98.2 %: m.p., 161-167°C. **(1)** was reacted with furan, giving 2-(4-nitrophenyl)furan **(2)** in a yield of 63.2 %: m.p. 132-132.9°C. **(2)** was converted to bis(4-(2-furyl)phenyl)diazene **(3)** with 89.9 % yield: m.p. 249-251°C; ^{1}H NMR (δ, $CDCl_3$), 8.3 (dd, 4H, Ph-*H*), 7.75 (d, 4H, Ph-*H*), 7.5 (m, 2H, furyl-*H*), 6.7 (m, 2H, furyl-*H*), 6.5 (m, 2H, furyl-*H*). **(3)** was hydrogenated, producing **(4)** in a yield of 98 %: m.p. 156-157°C; ^{1}H NMR (δ, $CDCl_3$), 7.53 (d, 4H, Ph-*H*), 7.39 (dd, 2H, furyl-*H*), 6.86 (d, 4H, Ph-*H*), 6.4 (m, 4H, furyl-*H*), 5.7 (s, 2H, N*H*). **(4)** was rearranged to the diamine (PFDA): mass/e 316 (M^+); m.p. 203-205°C; ^{1}H NMR (δ, acetone-d_6), 7.5 (d, 4H, Ph-*H*), 6.72 (d, 4H, Ph-*H*), 6.68 (d, 2H, furyl-*H*), 6.62 (m, 2H, furyl-*H*), 4.88 (s, 4H,NH_2). Elemental anal. Calcd. for $C_{20}H_{16}N_2$: C, 75.93; H, 5.10; N, 8.86. Found: C, 75.68; H, 5.02; N,

9.04. Here, the product was obtained with a yield of 20 %.

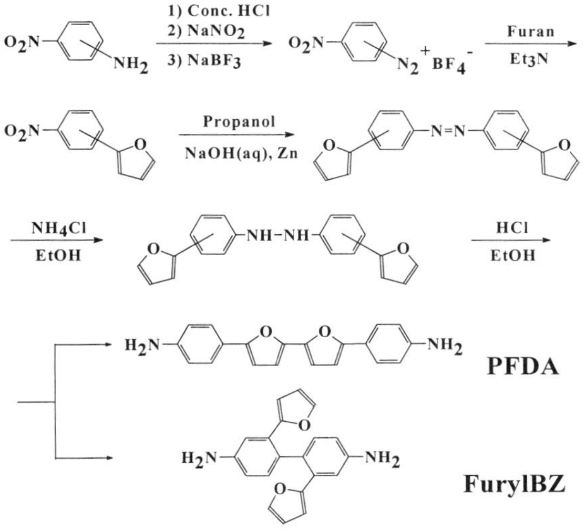

PFDA

FurylBZ

Fig. 1: Synthetic scheme of new blue-light-emitting monomers, 5,5'-bis(4-aminophenyl)-2,2'-bifuryl (PFDA) and 2,2'-bis(furyl)benzidine (FurylBZ).

2,2'-Bis(furyl)benzidine (FurylBZ) was also synthesized from 3-nitoaniline in the same manner as the PFDA was prepared: m.p. 156-157°C; ^{1}H NMR (δ, acetone-d_6), 7.37 (d, 2H, furyl-*H*), 7.25 (d, 2H, Ph-*H*), 6.84(d, 2H, Ph-*H*), 6.67 (dd, 2H, Ph-*H*), 6.17(dd, 2H, furyl-*H*), 5.42 (d, 2H, furyl-*H*), 4.78 (s, 4H,N*H*2). Elemental anal. Calcd. for $C_{20}H_{16}N_2$: C, 75.93; H, 5.09; N, 8.85. Found: C, 75.71; H, 4.87; N, 9.50.

In addition, another four diamine monomers were newly prepared as follows. 6-(4-Phenylphenoxy)hexanol was derived from 6-bromohexanol and 4-phenylphenol in accordance to a method described in the literature[2]. In the same manner, there were prepared 6-(3-phenylphenoxy)hexanol, 6-(2-phenylphenoxy)hexanol, and 6-(4-biphenylmethoxy)hexanol. These hexanol derivatives were substituted to 2,4-dinitrofluorobenzene by the etherification as reported previously in the literature[3]. The obtained dinitrobenzene derivatives were converted to the corresponding diamines: 2,4-diamino-(6-(4-phenylphenoxy)-hexyloxy)benzene (BP1), 2,4-diamino-(6-(3-phenylphenoxy)hexyloxy)benzene (BM1), 2,4-diamino-(6-(2-phenylphenoxy)hexyloxy)benzene (BO1), and 2,4-diamino-(6-(4-biphenylmethoxy)hexyloxy)benzene (BP2).

Synthesis of poly(amic acid)s and polyimides. The purified pyromellitic dianhydride (PMDA) of 1.3970 g (7.390 mmol) was added slowly to the PFDA of 2.3358 g (7.390 mmol) dissolved in 36 mL *N*-methyl-2-pyrrolidone (NMP) with stirring vigorously in a glove box filled with dry nitrogen gas (see Fig. 2). Once the dianhydride addition was completed, the reaction flask was capped tightly and stirring was continued for 1 day to make the polymerization mixture completely homogeneous, finally giving a poly(amic acid) (PAA) solution, PMDA-PFDA PAA. In the same manner, PMDA-FurylBZ PAA was prepared in NMP from PMDA and FurylBZ monomers.

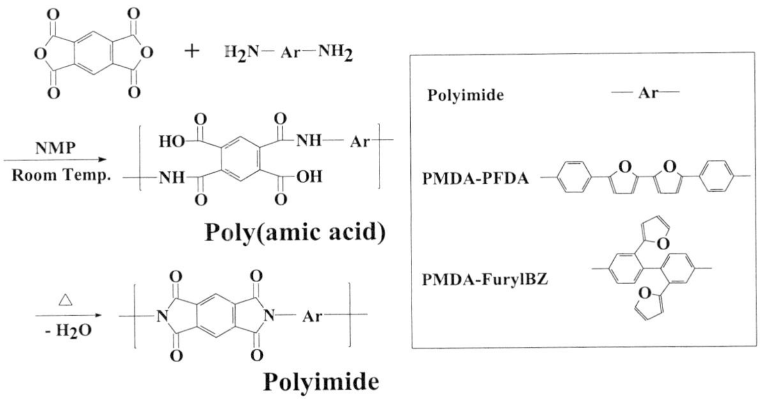

Fig. 2: Synthetic scheme of soluble poly(amic acid)s and polyimides from pyromellitic dianhydride (PMDA) with 5,5'-bis(4-aminophenyl)-2,2'-bifuryl (PFDA) and 2,2'-bis(furyl)-benzidine (FurylBZ).

Another series of PAA precursors were synthesized in NMP by the polycondensations of 4,4'-oxydiphthalic anhydride (ODPA) with *m*-phenylene diamine (MPD) and its derivatives containing short side chains, such as BP1, BM1, BO1, and BP2: ODPA-MPD PAA, ODPA-BP1, ODPA-BM1 PAA, ODPA-BO1 PAA, and ODPA-BP2 PAA (see Fig. 3).

Solid contents of the PAA precursor solutions were 8-12 wt.%. These solutions were filtered with silver metal membranes of 1.0 μm pore size, tightly sealed, and stored in a refrigerator of -4°C before use. For these precursor polymers, the intrinsic viscosity [η] was measured in NMP at 25.0°C using an Ubbelohde suspended level capillary viscometer as a method described in the literature[4,5]: 0.661 dL/g for PMDA-PFDA PAA, 0.545 dL/g for PMDA-FurylBZ PAA, 0.207 dL/g for the ODPA-BP1 PAA, 0.259 dL/g for the ODPA-BM1 PAA, 0.653 dL/g for the ODPA-BO1 PAA, 0.245 dL/g for the ODPA-BP2 PAA, and 0.400 dL/g for the ODPA-MPD PAA.

The PAA solutions were spin-coated on glass substrates, followed by soft-baking on a hot plate at 80°C for 1 h. The soft-baked precursor films were thermally imidized in an oven with a dried nitrogen gas flow by a three-step protocol: 150°C/30 min, 230°C/30 min, and 300°C/60 min. A ramping rate was 2.0 K/min for the first two steps and 5.0 K/min for the final step. After imidization, the samples were cooled to room temperature with a rate of 10.0 K/min. The thickness of the obtained polyimide (PI) films was in the range of 3-20 μm. The films were taken off from the glass substrates with the aid of deionized water, followed by drying for 2 days at 50°C in a vacuum of 5×10^{-4} torr. For measurements of physical properties, films were cut into appropriate sizes. In addition, for uv-visible spectroscopic measurements, polyimide films of 0.1-2 μm thickness were prepared on quartz substrates from a diluted solution (1-2 wt.% solid content) of the poly(amic acid).

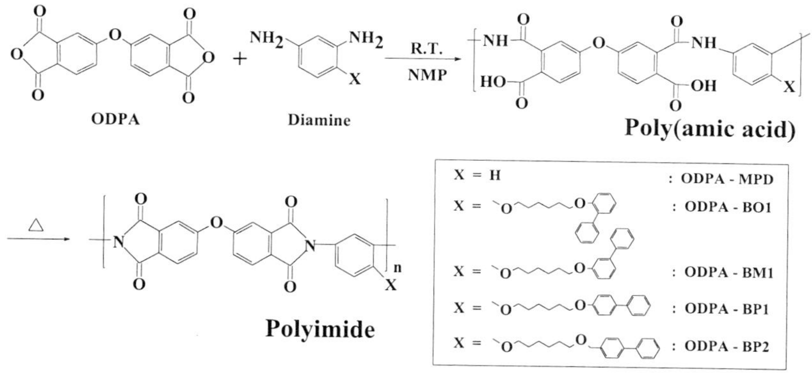

Fig. 3: Synthetic scheme of poly(amic acid)s and polyimides from ODPA and diamines with various side chains.

Characterization. The product obtained at each synthetic step was identified by ^{1}H NMR spectroscopy. ^{1}H NMR spectra were measured using a Bruker ASPECT 300 MHz NMR spectrometer in acetone-d_6 or chloroform-d_1 solvent. Melting points were determined using a Thomas Hoover capillary melting point apparatus. In addition, for the newly synthesized diamine monomer, the molecular weight was determined using a mass spectrometer (Hewlett-Packard HP-5970) coupled with a gas chromatographic analyzer (Hewlett-Packard HP-5890). Elemental analysis was carried out using an Elemental Vario analyzer.

Thermogravimetric analysis (TGA) was carried out for the soft-baked poly(amic acid)s and their polyimide using a Perkin-Elmer thermogravimeter (Model TGA7) under a dry nitrogen gas flow. That is, a baked precursor sample was heated up to 300°C with a rate of 5.0 K/min, then soaked at 300°C for 1 h, and finally cooled to 50°C. Then, the imidized sample was again heated up to 800°C with a ramping rate of 5.0 K/min.

Glass transition temperatures (T_g's) were measured over 25-400°C using a Seiko differential scanning calorimeter (Model DSC-220CU). During the measurement, dry nitrogen gas was purged with a flow rate of 80 cc/min and a ramping rate of 10.0 K/min was employed.

In scanning run, a sample of about 5 mg was used. The temperature calibration was carried out using indium and tin, and the enthalpy calibration using indium. T_g was taken as the onset temperature of glass transition. In-plane thermal expansion coefficients (TECs) were measured with a load of 5 g at a ramping rate of 5.0 K/min over the range of 50-150°C under a nitrogen gas flow using a Seiko thermomechanical analyzer (Model TMA-120 CU) with a tension probe. The gauge length was 20.0 mm and the width of film strips was 2 mm. For each film strip, the measured TECs were averaged out over the range of 70-120°C.

For the new PFDA and FurylBZ monomers and their polyimides, excitation and photoluminescence spectra were measured at room temperature using a fluorescence spectrophotometer (Shimadzu RF-500) with a Xenon lamp. For both excitation and emission monochromators, bandpasses were 2 nm, respectively. Photoluminescence behavior of the new monomer was investigated for its solution with a concentration of 1.5 x 10^{-5} g/mL in 1,4-dioxane. The excitation wavelength employed was 340 nm. On the other hand, the polyimide in films was excited with a wavelength of 380 nm in a front-face arrangement in order to minimize the self-absorption. The size of polyimide specimens was 10 x 30 mm. UV-visible spectroscopic measurements were performed using a Hewlett-Packard spectrometer (Model HP-8452A).

In addition, relative fluorescence quantum yield (Φ_f) of the diamine monomer was measured in 1,4-dioxane by comparing the ratio of fluorescence emission intensity maximum to uv absorbance at the excitation wavelength used for the sample with that of a standard, quinine sulfate[6-9)]. For the polyimide in films, Φ_f was determined by the comparison with *N,N'*-dicyclohexylperylenebisimide dye dispersed in poly(methyl methacrylate)[10)].

Some of polyimide films coated on substrates were rubbed with varying rubbing density (L/l)[11,12)] using a laboratory rubbing machine with a roller covered by a rayon cloth (Yoshikawa Company, YA-20-R): $L/l = N[(2\pi rn/60v) - 1]$ where L is the total length (mm) of the rubbing cloth which contacts a certain point of the polymer film, l is the contact length (mm) of circumference of rubbing roller, N is the cumulative number of rubbings, n is the speed (rpm) of rubbing roller, r is the radius (cm) of roller, and v is the velocity (cm/s) of the substrate stage. Here, the thickness of polyimide films used as alignment layers was in the range of 4.0-5.0 μm.

Two pieces of the rubbed substrates were assembled together in the anti-parallel rubbing direction by using 50 μm thick poly(ethylene terephthalate) film spacers. A nematic LC, 4-*n*-pentyl-4'-cyanobiphenyl (5CB) was injected into the cell gap, followed by sealing the injection hole with an epoxy glue. Then, pretilt angle (α) of the LC molecules was measured using a crystal rotation apparatus which was made in our laboratory[12)].

Results and Discussion

Synthesis

PFDA and FurylBZ monomers were synthesized from 4-nitroaniline and 3-nitroaniline, respecively, by the five-step reaction sequence as shown in Fig. 1. Depending on the nitroaniline isomers, the reaction yield was 97-99 % at the first-step, 63-64 % at the second-step, 55-90 % at the third-step, 98 % at the fourth-step, and 20-21 % at the fifth-step. In comparison, the final step gave the lowest product yield. The low yield was attributed to possibly many side reactions occurred in the benzidine rearrangement reaction.

Phenylphenoxyhexanol isomers were synthesized with yields of 45-59 % from the reactions of 6-bromohexanol with three isomeric phenylphenols. In addition, 6-(4-biphenylmethoxy)hexanol was prepared with a yield of 53-54 % from 6-bromohexanol and 4-biphenylmethanol. Here, before reacting with phenylphenol isomers, 6-bromohexanol was protected with 3,4-dihydro-2*H*-pyran which could be deprotected in mild acidic condition, in order to avoid the reaction between its hydroxy and bromide group. The obtained phenylphenoxyhexanol isomers were further reacted with 2,4-dinitrofluorobenzene, respectively, leading to 2,4-dinitrobenzenes with four different side chain analogs consisting of a flexible spacer and a biphenyl mesogen. Yields of these coupling reactions were 89-92 %. The obtained dinitrobenzene derivatives were reduced to the corresponding diamine monomers: the yields were 42-47 %. These low yields might be due to dissociation reactions of the ether linkages in the relatively strong acid condition. However, the diamine products were successfully purified by column chromatography with silica gel. Through the whole reaction steps, the total yield was 18-25 %, depending on the diamine derivatives.

Using the newly synthesized diamine monomers, various poly(amic acid)s were prepared in NMP by the conventional low temperature polycondensations with PMDA and ODPA monomers. All the soluble poly(amic acid)s were obtained with reasonably high molecular weights: the intrinsic viscosity [η] was in the range of 0.400-0.661 dL/g, depending upon the poly(amic acid)s. Some of PAA precursors had relatively low [η] values (0.207-0.259 dL/g) but gave a good quality of films in the formation of polyimide films.

Imidization and thermal properties

The prepared poly(amic acid) solutions were spin-cast on glass substrates, followed by soft-baking at 80°C. For the soft-baked PAA films, thermal imidization was performed under a dry nitrogen atmosphere by thermogravimetry (TGA). In the heating run, the weight dropped sharply over 100-186°C, which is due to the removal of residual NMP solvent as well as the byproduct (that is water) of imidization in part. The further weight loss occurred over 186-300°C, due to the removal of the byproduct as well as the continuous removal of residual solvent. Then, the weight loss levels off during soaking for 1 h at the final imidization temperature, 300°C.

For polyimides prepared by imidization at 300°C, thermal stability was examined under a dry nitrogen atmosphere by TGA. The PMDA-PFDA PI film was stable up to ca. 370°C and then gradually degraded until ca. 520°C. Thereafter, the rate of the weight loss is significantly increased with increasing temperature. A similar TGA thermogram was obtained for the PMDA-FurylBZ PI. However, the polymer was stable up to ca. 400°C.

The ODPA-MPD PI without any side group was stable up to 480°C and began to degrade with increasing temperature. The polymer shows a small weight loss over the range of 480-620°C and then a large weight loss above 620°C: that is, the polyimide apparently reveals a two-step weight loss behavior. However, the suggested two-step degradation mode needs to be further investigated in details for full understanding. In comparison, the side chain containing polyimides show distinct two-step weight loss behaviors: the weight loss at the first-step might be due to the degradation of side groups and that at the second-step due to the degradation of polyimide main chain. In conclusion, the incorporation of side groups reduces the thermal stability of ODPA-MPD PI. In comparison, the thermal stability is in the decreasing order ODPA-MPD PI (T_d = ca. 480°C) > ODPA-BP1 PI (ca. 400°C) ≈ ODPA-BM1 PI ≈ ODPA-BO1 PI > ODPA-BP2 (ca. 330°C): here, T_d is the onset temperature of degradation.

In addition, in-plane thermal expansion coefficient (TEC) and glass transition temperature (T_g) were measured for the polyimides. The ODPA-MPD PI without side chain exhibited 33 ppm/°C TEC and 271.4°C T_g. However, all the side chain containing polyimides revealed relatively high TECs and low T_gs. TEC and T_g were 68 ppm/°C and 150.4°C for ODPA-BP1, 65 ppm/°C and 140.5°C for ODPA-BM1, 69 ppm/°C and 139.3°C for ODPA-BO1, and 58 ppm/°C and 148.5°C for ODBP2 PI, respectively. Conclusively, both TEC and T_g of ODPA-MPD PI are reduced by incorporating side chains. For the side chain containing polyimides considered, their TEC varies further with the flexible spacer length and its conformation, whereas the T_g changes further with the isomeric structure of biphenyl group.

Light-emitting characteristics

For the PFDA and FurylBZ solutions in 1,4-dioxane (1.5 x 10^{-5} g/mL), the uv-visible spectra exhibited broad featureless absorptions with the lowest energy absorption maximum at 376 nm and 326 nm, respectively. These absorptions are considered to be the π-π^* transition, leading to the formation of singlet excitons. With this UV-visible absorbance characteristic, the monomer solutions were excited at a wavelength of 340 nm, which provide the best fluorescence spectra. The solutions exhibited a strong emission peak centered at 442-443 nm, regardless of the isomeric structures. The relative fluorescence quantum yield (Φ_f) was measured to be Φ_f = 92 % for the PFDA and Φ_f = 52 % for the FurylBZ. Conclusively, these new diamine compounds emit an intense blue fluorescence.

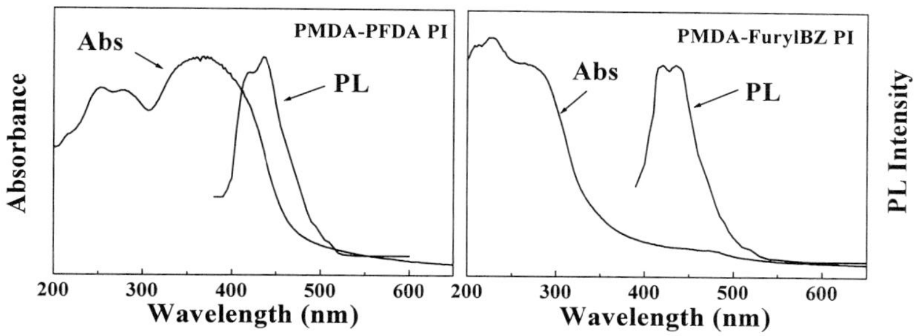

Fig. 4: UV-visible absorption (Abs) and photoluminescence (PL) spectra of the polyimide films, which were imidized at 300°C for 1 h: the film thickness was 0.6 μm for the measurement of UV-visible spectrum and 15.5 μm for the measurement of PL spectrum. The excitation was at 380 nm.

On the other hand, the PMDA-PFDA PI film exhibited a very broad featureless absorption at < 470 nm, while the PMDA-PFDA PI film showed a very broad featureless absorption at < 400 nm (see Fig. 4). When the PI films were excited at 380 nm, the best emission spectra were obtained. As shown in Fig. 4, the measured emission spectra are very similar to those of the PFDA and FurylBZ solutions, respectively. However, each PI film showed an emission peak having two peak maxima rather than a single peak maximum: 419 nm and 436 nm for the PMDA-PFDA PI; 419 nm and 435 nm for the PMDA-FurylBz PI. The two peaks in the emission spectrum of each PI film may be a vibronic structure instead of two separate peaks. In comparison, the emission spectrum of each PI film is relatively broader than that of its diamine monomer. This may result mainly from the nature of vibronic structure in the emission spectrum. However, the emission spectra of these PI films still are relatively much narrower than those of other light-emitting polymers, such as poly(*p*-phenylenevinylene) and its copolymers[1,6,13-16]. The Φ_f was measured to be 1.2 % for the PMDA-PFDA PI and 7.4 % for PMDA-FurylBZ PI. The Φ_fs are much higher than that (9.7×10^{-7}) of poly(4,4'-oxydiphenylene pyromellitimide)[17] which is the representative of conventional polyimides. In particular, the Φ_f of PMDA-FurylBZ PI is comparable to that (0.085-0.115) of poly(2-methoxy-5-(2'-ethyl)hexyloxy-*p*-phenylenevinylene)[18] (that is, an analog of PPV) which emits orange light. However, the PMDA-PFDA PI has a relatively low Φ_f value, compared to the PMDA-FurylBZ PI. This might be attributed to the ordered structure in the PMDA-PFDA PI film, which is evident in the X-ray diffraction pattern.

Alignment of LC molecules

The polyimide films were rubbed with varying rubbing density. For the rubbed polyimide films, the formation of microgrooves on the surface was confirmed to be in the rubbing direction by atomic force microscopy. The rubbed polyimide films were used to make LC cells as alignment layers. For the LC cells filled with 5CB molecules which are a nematic LC, the pretilt angle (α) of LCs was measured by the crystal rotation technique as a function of rubbing density. The results are presented in Fig. 5.

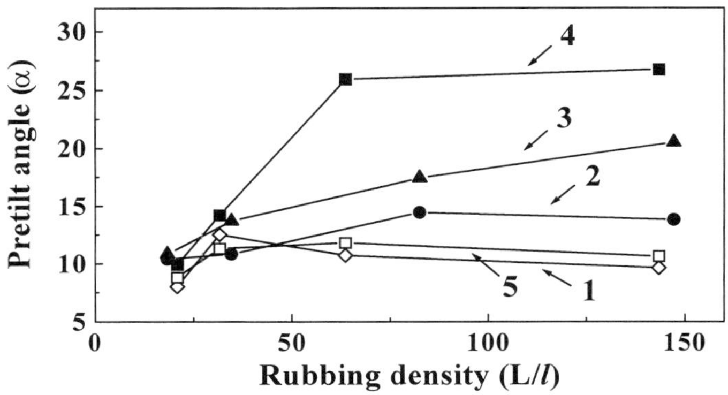

Fig. 5: Variations of the pretilt angle of LC (5CB) molecules in contact with the rubbed surfaces of polyimide films as a function of rubbing density (L/*l*): 1, ODPA-BP2 PI; 2, ODPA-BO1 PI; 3, ODPA-BM1 PI; 4, ODPA-BP1 PI; 5, ODPA-MPD PI.

For the LC cell fabricated with the ODPA-MPD PI film, the α of LC molecules increased initially with increasing rubbing density, but leveled off at ca. 31.5 rubbing density. Otherwise, the α decreased slightly over the high rubbing density range of > 63.6. A maximum α of 11.8° was achieved at a rubbing density of 63.6. This α value is relatively larger than those (0.7-5.1°) achieved for the 5CB filled LC cells fabricated with various polyimides without side chains[12,19]. The α value is comparable to those (13.0-14.0°) for the LC cells fabricated by some other polyimides with and without long alkyl side chains[19]. Similar results were obtained for the LC cells prepared with the ODPA-BP2 PI and ODPA-BO1 PI. However, the α values were achieved in high level, compared to the ODPA-MPD PI.

Different from the PIs discussed above, both ODPA-BM1 PI and ODPA-BP1 PI gave very large α values in the LC cells. For the LC cell with ODPA-BM1 PI, the α increased continuously with increasing rubbing density over the range of rubbing density (< 150) considered in the present study: that is, the α value varied in the range of 10.8-20.5° over < 150 rubbing density. For the LC cell fabricated with ODPA-BP1 PI, the α increased with rubbing density and leveled off at 63.6 rubbing density. Thereafter, the α further increased very slowly with rubbing density. An extremely large α of 26.0-27.0° was achieved over $\geq$ 63.6 rubbing density. In comparison, an achievable maximum pretilt angle is in the increasing order ODPA-MPD $\leq$ ODPA-BP2 < ODPA-BO1 << ODPA-BM1 < ODPA-BP1.

Conclusions

New PFDA and FurylBZ monomers emitting highly intense blue lights were synthesized, and their polyimides were prepared by the polycondensation reactions with PMDA and subsequent thermal imidization. The monomers, as well as their polyimide films were characterized by photospectrometry. The polyimides in films also emitted blue lights as detected for the monomers. The polymers were thermally stable up to 370-400°C. Conclusively, these polymers are considered as potential candidate materials for fabricating optoelectronic devices including LEDs which can emit blue light.

MPD derivatives having three different isomeric 6-(phenylphenoxy)hexyloxy groups as well as 6-(4-biphenylmethoxy)hexyloxy group as side chains were newly synthesized. Using these monomers, a series of polymers were synthesized by the polycondensations with ODPA and subsequent thermal imidization. These polymers were thermally stable up to 330-480°C, depending upon the side groups as well as the length of flexible spacers in the side group. In addition, T_gs and TECs were determined: by incorporating side chains, the TEC was increased and the T_g was decreased.

In addition, all the polyimides exhibited excellent performance in the rubbing process and the controlling of the alignment of LC molecules in the LC cell. The rubbing process with a relatively low rubbing density of 30-85 provided these polyimides an excellent controllability to pretilt LC molecules properly and to align LC molecules highly in the rubbing direction, which are highly desired in the LCD industry. The LC molecules were aligned well in the rubbing direction even by rubbing with a low rubbing density. The pretilt angle of LC molecules was achieved in a wide angle range of 8-27°, depending upon the rubbing density as well as the sort of polyimides. The pretilting behavior of LC molecules on the rubbed surface were very sensitive to all the molecular parameters in addition to the rubbing process. That is, a high flexibility of and a highly random orientation of polymer chains cause the LC molecules to pretilt with a relatively high angle. And, the conformation of flexible spacer and the isomeric structure of biphenyl mesogen end group in the side chain involve highly in controlling the pretilt of LC molecules. For the side chain containing polyimides studied here, the 6-(phenylphenoxy)hexyloxy and 6-(4-biphenylmethoxy)hexyloxy side chains, which are much shorter in length than the long alkyl side chains in the polyimides being used widely as LC alignment layers, were evident to involve effectively in the alignment of and the pretilt of LC molecules which are highly desired in the LCD industry. In particular, the major role of these side chains in pretilting LC molecules in a proper way might be attributed mainly to a strong interaction between the biphenyl mesogen end group and the mesogen unit of LC molecules. In conclusion, all the short side branched polymers have good properties suitable for applications in LCDs as well as microelectronic devices.

Acknowledgments
This study was supported in part by the KOSEF (ERC-CafPoly), by the Ministry of Education (New Materials Program in 1997), and by the Ministry of Industry & Energy and the Ministry of Science & Technology (MOST) (Electronic Display Industrial Research Association - G7 Project Program).

References

1) J. H. Burroughes et al., *Nature* **347**, 539 (1990)
2) N. Miyashita, A. Yoshikoshi, and P. A. Grieco, *J. Org. Chem.*, **42**, 3772 (1977)
3) T. Ichino et al., *J. Polym. Sci.: Part A: Polym. Chem.*, **28**, 323 (1990)
4) S. I. Kim, S. M. Pyo, K. Kim, M. Ree, *Polymer* **39**, 6489 (1998)
5) S. I. Kim, S. M. Pyo, M. Ree, *Macromolecules* **30**, 7890 (1997)
6) A. Hilberer, H.-J. Brouwer, B.-J. V. D. Scheer, J. Wildeman, G. Hadziioannou, *Macromolecules* **28**, *28*, 4525 (1995)
7) J. C. Phelan, C. S. P. Sung, *Macaromolecules* **30**, 6837 (1997)
8) J. N. Demas, G. A. Crosby, *J. Phys. Chem.* **75**, 991 (1971)
9) W. H. Melhuish, *J. Phys. Chem.* **65**, 229 (1961)
10) M. Hasegawa, Y. Shindo, T. Sugimura, S. Ohshima, K. Horie, M. Kochi, R. Yokota, I. Mita, *J. Polym. Sci.: Part B: Polym. Phys.* **31**, 1617 (1993)
11) T. Uchida, M. Hirano, and H. Sakai, *Liquid Crystal*, **5**, 1127 (1989)
12) S. I. Kim, S. M. Pyo, M. Ree, M. Park, and Y. Kim, *Mol. Cryst. Liq. Cryst.*, **316**, 209 (1998)
13) Z. Yang, I. Sokolik, F. E. Karasz, *Macromolecules* **26**, 1188 (1993)
14) Z. Yang, F. E. Karasz, H. J. Geise, *Macromolecules* **26**, *26*, 6570 (1993)
15) D.-H. Hwang et al., *Bulln. Korean. Chem. Soc.* **16**, 35 (1995)
16) T. Zyung et al., *Chem. Mater.* **7**, 1499 (1995)
17) G. Arjavalingam, G. Hougham, J. P. Lafemina, *Polymer* **31**, 840 (1990)
18) J. C. de Mello, H. F. Wittmann, R. H. Friend, *Adv. Materials* **9**, 230 (1997)
19) K.-W. Lee, S.-H. Paek, A. Lien, C. During, and H. Fukuro, *Macromolecules*, **29**, 8894 (1996)

Fluorinated Heteroaromatic Polyethers for Low Dielectric Constant / High Temperature Applications

Gerhard Maier*, Jörg Haußmann, Marcus Dietlmeier, Susanta Banerjee

Lehrstuhl für Makromolekulare Stoffe, Technische Universität München, Lichtenbergstraße 4, D-85747 Garching, Germany

SUMMARY: Several series of poly(arylene ether)s with trifluoromethyl substituents were prepared and characterized. These materials are potential candidates for the use as low dielectric constant insulators (intermetal dielectrics, IMD, and interlayer dielectrics, ILD) on microchips. Thermal stability up to 450 °C and a dielectric constant below 3 preferably below 2.5) is required for this application. The thermal stability of the poly(arylene ether)s was increased from 320°C to more than 500 °C by optimization of the structure of the repeating unit. The dielectric constant of one of the most promising structures was determined to be 2.8. In addition, plasma polymerized thin films from hexafluorobenzene, tetrafluorobenzene, perfluorotoluene and perfluorodecaline were prepared and characterized with respect to solubility, dielectric constant, adhesion, and thermal stability.

Introduction

The continuing trend to increase the capacity of memory chips in microelectronics and the resulting need to decrease of the size of the devices (transistors, capacitors etc.) on the chips has reached a stage where new materials are required for the conducting interconnection layers on the chips[1,2,3,4]. Driving force for the fabrication of chips with smaller structures are the decreasing cost per device, reduced power consumption, and decreasing intrinsic gate delay of smaller transistors. The major problems caused by this trend, however, are the decreasing speed of signal propagation, caused by increasing capacitance of the interconnecting metal lines, and increased cross talk as the distance between the metal lines shrinks. The signal delay time τ depends on the resistance and capacitance of the metal lines[5]:

$$\tau = RC = 2\rho\epsilon\epsilon_0\left[\frac{4L^2}{P^2} + \frac{L^2}{T^2}\right]$$

τ: signal delay time
R: resistance
C: capacitance
ρ: specific resistance of the conductor
ϵ: dielectric constant of the insulating material
ϵ_0: dielectric constant of vacuum
L: length of the conductor
T: thickness of the conductor
P: distance between two conducting lines

 CCC 1022-1360/99/$ 17.50+.50/0

Figure 1 shows the signal delay time in dependence of the size of the smallest structures on a chip for a given layout of the interconnect lines.

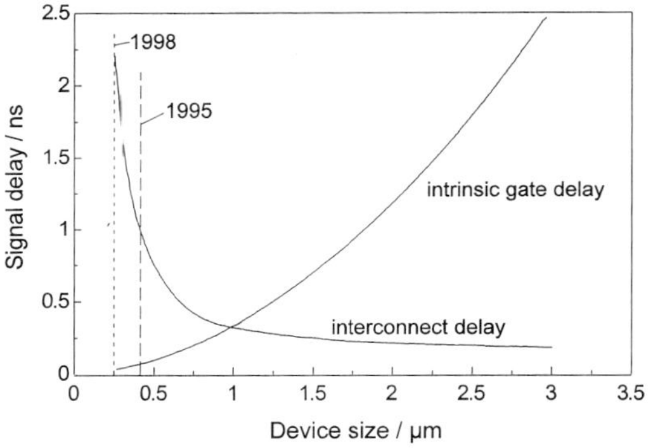

Figure 1: Signal delay time vs. device size for a given layout of the interconnect lines[1,6)]

The graph in Figure 1 clearly shows that at feature sizes below 1 µm the signal delay time is dominated by the interconnect delay rather than the intrinsic gate delay. At sizes of 0.25 µm as used typically for DRAM memory chips and CPU's commercialized initially in 1998[3,6)], the signal delay is approximately 2 ns, limiting the maximum clock speed of these chips to 500 MHz. Further acceleration requires changes of the layout, use of an interconnect metal with higher conductivity than the presently used aluminum based alloy, or the use of an insulating material with a lower dielectric constant than the presently used[1,2,4,7,8)] SiO_2 with $\epsilon \geq 4$. Changes of the layout are generally strongly restricted by the desire to keep the chip size as small as possible and the limits of available photolithographic techniques. This can be compensated in part by extending the interconnects into the third dimension, that is by adding additional metallization layers. At present, typically 3-4 metallization layers are used, while the 4 GB RAM chips are expected to have up to 6-8 layers. In 1998, copper has been introduced as interconnect metal to replace the aluminum based alloy. This increases the signal speed by approximately 35% due to the higher specific conductivity of copper[7)]. However, copper metallization required a considerable change in processing, since dry etching and gas phase deposition of copper is difficult and slow. This new „copper damascene“ process[2,7)] is too complex to be described here in detail. In principle, in this process the insulating material, e.g. a polymer, is deposited first, then trenches are formed in this layer by photolithographic methods, the trenches are then filled with copper by a combination of gas phase and galvanic deposition techniques, and then the excess copper is removed by chemical mechanical polishing (CMP). Etch stop layers, polish stop layers, and diffusion barriers for the copper as well as adhesion promoting layers are required to obtain reliable structures.

A quite promising concept to reduce the signal propagation time is the use of insulating materials with dielectric constants below that of SiO_2. Polymers can exhibit dielectric constants down to $\epsilon \approx 2$ for aliphatic perfluorinated Materials such as PTFE or amorphous TFE/Perfluorocycloolefin-co-polymers, the lowest value known for any dense, non-porous material[9)].

REQUIREMENTS FOR LOW-ϵ-MATERIALS

The need for incorporation of the new low-ϵ-materials into existing schemes for manufacturing of microchips results in strict requirements regarding various properties. The most important issues[4)] are low dielectric constant, high temperature resistance, low moisture absorption, and good adhesion (Table 1).

Table 1: Requirements for low-ϵ-materials[4)]

Property	ϵ	T_d	moisture uptake	adhesion to
Value	≤ 2.5	>450°C	< 1%	Al, Si, SiO_2, Cu, ...

At present, a dielectric constant of 2.5 or below is targeted, which corresponds to a decrease of the signal delay τ to 62% compared to a structure with SiO_2 as insulating layer. The processing temperature will be at least 400 °C in each metallization step, even in the copper metallization scheme[7,4)]. In addition, there are many other parameters mostly concerning mechanical properties, coefficients of thermal expansion, compatibility with existing etching processes, compatibility with chemical mechanical polishing, planarization, etc[4)].

In order to keep the dielectric constant of a polymer low, polar and polarizable structures have to be avoided[9)]. This also helps to decrease moisture uptake. It has been found that the incorporation of fluorine also decreases the dielectric constant[9)]. Bulky, rigid structural elements increase the free volume of a polymer, which also results in a decreased value of ϵ[9)]. A large number of different polymer classes has been suggested for the use as low-ϵ-materials. The list[4,1,2,8,7)] includes polyimides, poly(benzoxazole)s, poly(quinoline)s, poly(aryl ethers) (each of these fluorinated or non-fluorinated), poly(silsesquioxane), poly(naphthalene), poly(perfluorocyclobutane), Parylene and Parylene F, PTFE and amorphous PTFE-copolymers. All of these polymers have their strengths and weaknesses, which have been discussed elsewhere[10)] in detail. Some of these polymers (poly(naphthalene), poly(perfluorocyclobutane), and the Parylenes) are prepared by chemical vapor deposition (CVD), plasma polymerization or a related technique directly on the substrate. The others are applied from solution by spin coating[4,1,2,8,7].

We are pursuing both the development of polymers for spin coating on the basis of fluorinated poly(aryl ethers) as well as the application of plasma polymerization for the preparation of fluorinated dielectric films.

Low-ϵ-poly(aryl ether)s for application from solution

Poly(aryl ether oxazole)s and poly(aryl ether thiazole)s

Poly(aryl ether)s are usually prepared by nucleophilic displacement of halogen from activated aryl bishalides by bisphenoxides[11,12,13]. The activation is achieved by electron withdrawing groups. Most of these groups are polar, such as carbonyl[13] and sulfone[12]. However, for the application discussed here the use of such functional groups is detrimental[9]. Activation can also be achieved by electron deficient heteroaromatic rings[11,14]. One of the least polar electron poor heteroaromatic groups is the oxazole ring[14,15]. Another ideally suited activating group is the perfluoroalkyl group[16,17], especially the trifluoromethyl group. Thus, polymers of the following type **2**[18,19], derived from monomer **1** appeared to be promising candidates (Scheme 1):

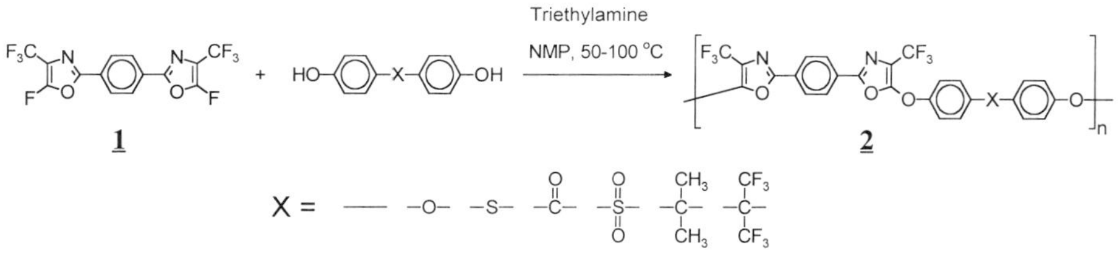

Scheme 1: Bisoxazole monomer **1** and the corresponding poly(aryl ether)s **2**

However, these polymers exhibit an unexpectedly low thermal stability with decomposition temperatures (initial weight loss) at 310 °C in air or nitrogen[18], independent of the exact structure of the repeating unit. The substitution pattern of the oxazole rings with a substituent in 4-position was identified to be the problem[18,19]. The CF_3-group stabilizes an intermediate formed in a thermal ring scission[18,19].

Consequently, similar polymers **4** with thiazole rings in place of the oxazoles were prepared (Scheme 2)[20]. Sulfur containing heteroaromatic rings are considered to be more stable due to their more pronounced aromatic character (higher number of mesomeric structures) than their oxygen analogs[21,22,23]. The thermal stability of the polymers **4** was indeed found to be higher with decomposition temperatures of 420°C[20].

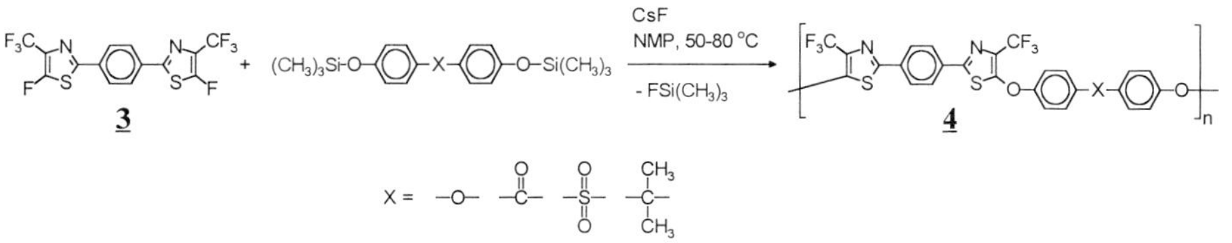

Scheme 2: Bisthiazole polymers **4**

In addition, mono-oxazole polymers **6** (Scheme 3) were studied[15]. In these polymers, there is no substituent in 4-position of the oxazole rings. They exhibit decomposition temperatures of the order of 450-490°C (10% weight loss in air)[15], which is typical for polymer with five-membered heteroaromatic rings in the main chain[14].

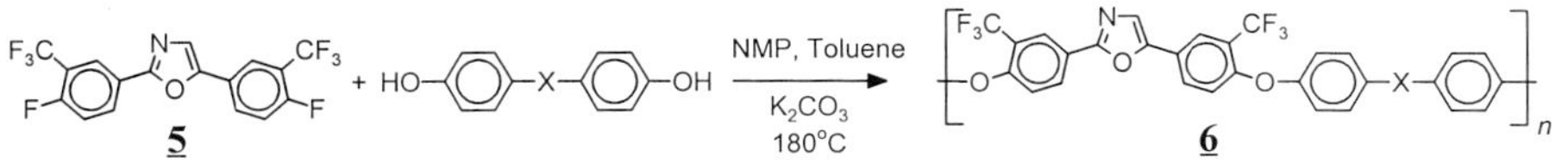

Scheme 3: Mono-oxazole polymers **6**

All polymers **2**, **4** and **6** are soluble in amidic solvents such as NMP, DMAc or DMF at room temperature[15,18,19,20]. Many of them are also soluble in chlorinated solvents and THF. Since the thermal stability of the polymers **6** appears to be sufficient, the dielectric constant of polymer **6a** with X = $-C(CF_3)_2-$ was determined. The presence of four trifluoromethyl groups per repeating unit results in a relatively low value of $\epsilon = 2.8$, which is a good basis for further developments.

The glass transition temperatures of the polymers **6** were found to be between 154 and 220°C (**6a**: 205°C)[15] and are lower than initially desired. The coefficient of thermal expansion increases above T_g. Thus, the T_g of polymers used as low-ϵ-dielectric material should be as close to the maximum processing temperature as possible, if not higher. The oxazole ring introduces a catenation angle[24] of approximately 132°, which increases the conformational flexibility of the polymer chain and hence decreases the glass transition temperature. Also, the oxazole ring is slightly polar with a dipole moment of $5 \cdot 10^{-30}$Cm[25] across the polymer chain. The replacement of the oxazole wing with a *para*-substituted phenyl ring was expected to decrease the dielectric constant and to increase the glass transition temperature at the same time.

Terphenyl-poly(aryl ether)s

Based on the considerations described above, we prepared terphenyl monomer 7, in which trifluoromethyl groups placed in *ortho*-position to fluorine atoms act as activating groups for the nucleophilic displacement reaction.

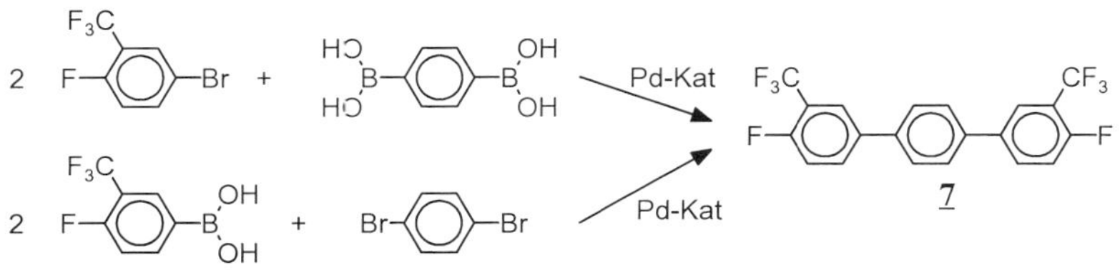

Scheme 4: Structure of terphenyl monomer 7

Monomer 7 was prepared by Suzuki-coupling[26,27] of either 2-fluoro-5-bromotrifluoromethyl benzene with benzene-1,4-bisboronic acid or 4-fluoro-3-trifluoromethyl benzene boronic acid with 1,4-dibromo benzene. The second approach proved to be much more successful with overall yields well above 80%, while yields were around 20% when the first pathway was followed. Possibly the trifluoromethyl group in the bromo-compound slows the coupling reaction to an extent were undesired side reactions become important.

The polycondensation of monomer 7 with various bisphenols under conventional reaction conditions proceeds quite successfully, and high molar mass polymers are obtained. Table 2 shows the structures of the resulting polymers **8** and their molar masses. The molar masses of the polymers **8** are quite high, with the exception of **8e**. While all other polymers **8** form very tough, flexible films, the films prepared from **8e** remain brittle. This can be attributed to the relatively low nucleophilicity of the phenoxide prepared from dihydroxydiphenyl sulfone, which is caused by the electron withdrawing effect of the sulfone group. Since the activation of the fluorine in monomer 7 is only due to the negative inductive effect of the trifluoromethyl groups in *ortho*-position rather than mesomeric stabilization of the intermediate Meisenheimer-Komplex, monomer 7 is not very reactive towards nucleophilic replacement of the fluorine atoms. Consequently, the substitution reaction is slow in the case of **8e**, and side reactions may become important, which limit the molar mass of the resulting polymer.

Table 2: Structures and molar masses of the polymers 8

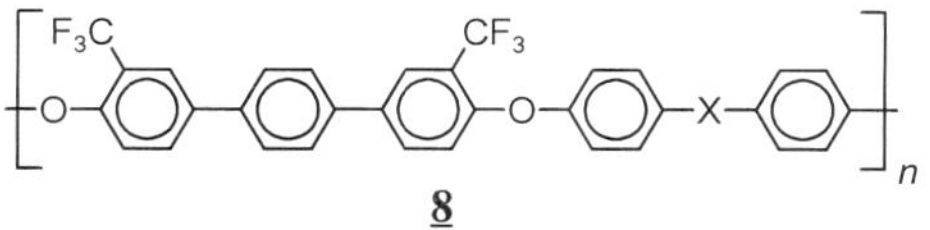

8

Polymer No.	X	$\bar{M}_n$ g · mol^{-1}	$\bar{M}_w$ g · mol^{-1}
8a	$-C(CH_3)_2-$	43 200	83 400
8b	$-C(CF_3)_2-$	30 000	74 500
8c	—	21 200	62 200
8d	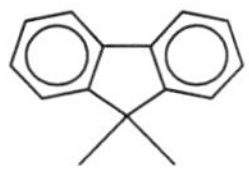	44 600	71 600
8e	$-SO_2-$	6 500	16 400

The decomposition temperatures of polymers 8c and 8d in air are well above 500 °C (expressed as 5% weight loss in dynamic thermogravimetry). However, the glass transition temperatures as determined by DSC and DMTA are lower than expected (Table 3).

Table 3: Thermal properties of the polymers 8

Polymer	X	Polymers 8 T_d [a)] °C	Polymers 8 T_g (DMTA)[c)] °C	Polymers 8 T_g (DSC)[b)] °C	Polymers 6 T_g (DSC) °C
8a	$-C(CH_3)_2-$	464	210	174	185
8b	$-C(CF_3)_2-$	500 (520)[d)]	218	197	205
8c	—	514	198	209	200
8d	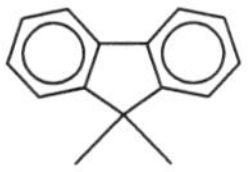	530	240	234	-
8e	$-SO_2-$	422	not determined	214	210

[a)] 5% weight loss, air, 10 K/min; [b)] 20K/min; [c)] 1 Hz, 2K/min; [d)] in nitrogen

The glass transition temperatures of the polymers 8 do not differ very much from those of the oxazole polymers 6 described above, despite the more rodlike terphenyl segment in the repeating unit of polymers 8. For 8a and 8b, there appears to be even a decrease in T_g as the oxazole ring is replaced by a 1,4-phenyl ring. Possibly the small dipole moment of the oxazole ring contributes to chain interactions which increase T_g, an effect which is not present when a phenyl ring replaces the oxazole. The polarity and geometry of the oxazole ring push T_g of the polymers 6 in opposite directions, and which of these influences dominates appears to depend on the rest of the repeating unit.

Some mechanical properties of the polymers 8 measured with thin film samples cast from THF solution are shown in Table 4.

Table 4: Mechanical properties of the polymers 8

Polymer	X	Young's Modulus[a)]	Tensile strength[b)]	Elongation at break[b)]
		GPa	MPa	%
8a	$-C(CH_3)_2-$	1.32	58	18
8b	$-C(CF_3)_2-$	2.51	55	28
8c	—	1.77	70	10
8d		2.69	61	7

[a)] DMTA, 1Hz, 2K/min; [b)] stress-strain measurements, room temperature, stain rate 5%/min

As expected, the polymer with the bulky fluorene group (**8d**) proved to be the most brittle one, but even the best (**8b**) is not very ductile, despite the relatively high molar masses of these polymers. Measurements of the dielectric constants are in progress.

Quadriphenyl-poly(aryl ether)s

All in all, the performance of the polymers 8 was not as good as expected. Therefore, another monomer with even more extended rodlike structure was prepared (9). The same synthesis strategy as for 7 was applied, with the exchange of 4,4'-dibromo biphenyl in place of 1,4-dibromo benzene in order to achieve a quadriphenyl unit:

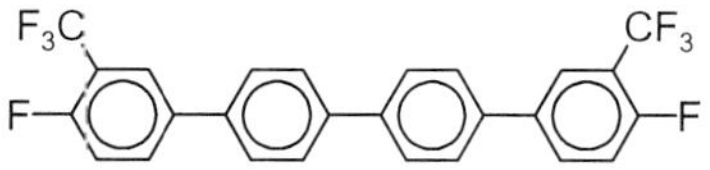

9

Scheme 5: Quadriphenyl monomer **9**

The reactivity of monomer 9 is very similar to monomer 7, and consequently under the same reaction conditions polymers 10 could be prepared with quite high molar masses (Table 5).

Table 5: Structures and molar masses of polymers 10

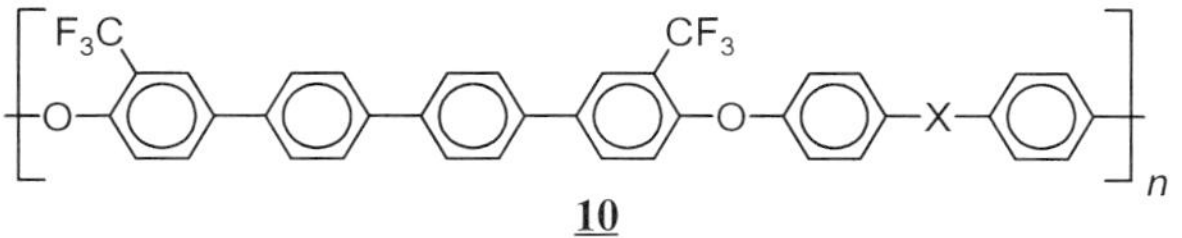

10

Polymer No.	X	$\bar{M}_n$ / $g \cdot mol^{-1}$	$\bar{M}_w$ / $g \cdot mol^{-1}$
10a	$-C(CF_3)_2-$	32 000	88 000
10b	—	39 300	94 400
10c	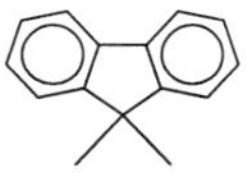	42 000	106 000

The thermal stabilities exceed 500 °C for all of these polymers, and the glass transition temperatures are considerably higher than those of the terphenyl polymers 8 (Table 6). Polymer 10c is especially promising with T_g of 300 °C and its high thermal stability. The structure of this polymer contains no polar groups except for the ether links. Also, the bulkiness of the fluorene unit can be expected to result in an increase in free volume. As a consequence, we expect this polymer to exhibit its dielectric constant to be considerably lower than that of polymer 6a, which was 2.8. The Measurements are in progress.

Table 6: Thermal stabilities and glass transition temperatures of polymers 10

Polymer	X	Polymers 10			Polymers 8
		T_d [a]	T_g (DMTA) [c]	T_g (DSC) [b]	T_g (DSC)
		°C	°C	°C	°C
10a	$-C(CF_3)_2-$	510	225	210	197
10b	—	530	268	257	209
10c	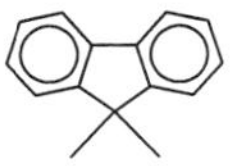	534	315	300	234

[a] 5% weight loss, air, 10 K/min; [b] 20K/min; [c] 1 Hz, 2K/min

In addition to the excellent thermal properties the mechanical properties of the polymers **10** are also quite encouraging. Dynamic-mechanical testing and stress-strain measurements of films revealed high strength and high drawability of these polymers. The data are summed up in Table 7. For comparison, typical values for commercial PEEK and Polycarbonate are added.

Table 7: Mechanical properties of the polymers **10**

Polymer	X	Young's Modulus[a)]	Tensile strength[b)]	Elongation at break[b)]
		GPa	MPa	%
10a	$-C(CF_3)_2-$	2.59	65	78
10b	—	2.04	75	76
10c	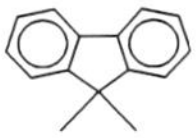	2.18	115	120
Victrex PEEK	-	3.6	95-100	>50
PC	-	2.4	72	120

a) DMTA, 1Hz, 2K/min; b) stress-strain measurements, room temperature, stain rate 5%/min

Tensile strength and elongation at break of polymer **10c** are comparable to typical engineering and high performance thermoplastic materials. The tensile modulus of **10c** is relatively low, but this may be due to the fact that we were using thin films cast from DMF solution with no optimization of solvent and casting conditions. The films were dried by heating and evacuation in several steps, with a final drying temperature of 170 °C under vacuum for 24 h and proved to be solvent free in dynamic thermogravimetry.

Fluorinated low-ϵ-films by plasma polymerization

Plasma polymerization proceeds by free radicals and/or diradicals produced from various precursors in the gas phase under reduced pressure by collisions of the precursor molecules with the atoms, ions and electrons of a plasma. The radicals are deposited on the substrate and combine to larger structures, and finally a polymer is formed. Depending on the precursor, mechanism and reaction conditions, the structure of these polymers is more or less defined, and often crosslinking occurs, causing insolubility of the deposited films. A schematic drawing of the plasma deposition chamber used for our experiments is shown in Scheme 6.

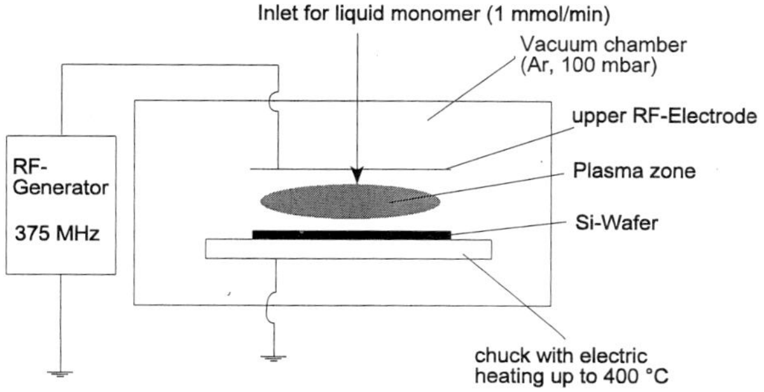

Scheme 6: Schematic drawing of the plasma deposition chamber

The need for high temperature stability of the deposited films caused us to use fluorinated aromatic compounds as precursors, because we expected higher thermal stability from aromatic or partially aromatic polymer films compared to purely aliphatic materials such as Teflon or Teflon AF. Therefore, we chose 1,2,4,5-tetrafluorobenzene, hexafluorobenzene[28)], and octafluoro toluene as precursors. Perfluorodecaline was also included in the studies, because it has a higher fluorine content than the aromatic compounds, which could result in a lower dielectric constant.

For the deposition experiments, argon at 100 mbar was used as carrier gas. The precursors were introduced at a constant rate of approximately 1 mmol/min, and the RF-generator was operating at 375 MHz at a power input of 20W. The substrate, a silicon wafer, was heated to various temperatures between 25 °C and 400 °C.

The first parameter studied in this series was the deposition rate in dependence of the substrate temperature. A minimum rate of 100 nm/min is desirable to sustain sufficient processing speed for chip production. Scheme 7 shows the results.

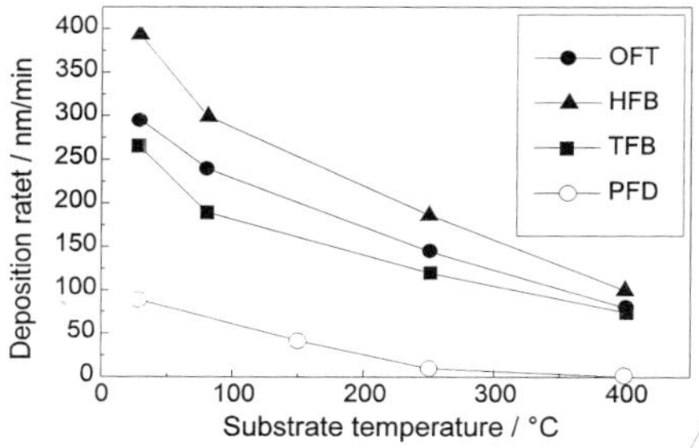

Scheme 7: Deposition rate at various substrate temperatures for:
octafluorotoluene (OFT), -●-; hexaflurobenzene (HFB), -▲-; tetrafluorobenzene (TFB), -■-; perfluorodecaline (PFD), -○-

All aromatic precursors tested here deposit sufficiently fast even at a substrate temperature of 400 °C. From perfluorodecaline no films are deposited at this high temperature. Even at 25 °C, film deposition from perfluorodecalin is slow under the conditions employed in our study. The thermal stability of the films from hexafluorobenzene and tetrafluorobenzene is acceptable. They show a weight loss of 3% at 400 °C within the first hour, but there is no further weight loss at this temperature after that. The film thickness remains constant during the initial weight loss, indicating that the weight loss is probably due to desorption of low molar mass absorbed oligomers rather than thermal decomposition of the films. Films deposited at lower substrate temperatures exhibit unsatisfactory thermal stability.

The adhesion of the films from hexafluorobenzene and tetrafluorobenzene also depends on the substrate temperature during deposition. While adhesion of films prepared as 25 °C is insufficient (tape test after scratching), adhesion to Si and Al of the films deposited at 400 °C is good. However, octafluorotoluene did not result in films with good adhesion under any conditions in the present study. All Films deposited at 400°C were insoluble in AZ thinner (a common solvent for photoresists, consisting of a mixture of aliphatic ethers), while some material could be extracted from those films prepared at 25°C from the aromatic precursors. Films deposited from perfluorodecaline were always insoluble. Internal stress in the films deposited at 400 °C is high. For hexafluorobenzene, it increases from 40 MPa for films deposited at 25°C to 190MPa for films deposited at 400°C. All films deposited in this study exhibited no detectable moisture absorption (« 1%).

The dielectric constants of the films deposited from the four precursors in this study are summarized in Table 8.

Table 8: Dielectric constants of the films prepared by plasma deposition

Precursor	Deposition temperature °C	Dielectric constant at 100 kHz
Perfluorodecalin	25	2.5-2.7
	400	no film
Octafluorotoluene	25	2.6-2.8
	400	not determined
Tetrafluorobenzene	25	2.7-2.9
	400	3.0-3.2
Hexafluorobenzene	25	2.8-3.0
	400	3.0-3.2

As can be seen from Table 8, the dielectric constants of the films prepared in this study are not as low as expected initially. IR spectra of the films reveal the presence of carbonyl functions in the deposited films. These are present even if the most stringent inert gas conditions are used during deposition. At present, in our opinion the presence of oxygen in the films is caused by reaction of oxygen with trapped free radicals during handling outside of the deposition chamber. So far, we have not succeeded in developing a deposition procedure which limits the number of trapped radicals to a value where this reaction with oxygen is no longer important.

Conclusion

Polymers with potential use as high temperature low-ϵ-dielectric materials for spin coating as well as plasma deposition were studied. Starting from a series of heteroaromatic poly(aryl ether)s with oxazole rings and trifluoromethyl substituents, the thermal stability and glass transition temperatures could be increased by changing the chemical structure of the repeating unit. The introduction of terphenyl groups into the polymer backbone did result in the expected improvement of the decomposition temperature, but it did not result in the expected increase in glass transition temperature. Only when quadriphenyl groups were used instead of the terphenyl units high T_g's of the order of 300 °C were achieved. These materials also exhibit excellent mechanical properties. Measurements of the dielectric constants are in progress. Based on the chemical structure, the prospects are promising because of the absence of polar groups (besides the ether bridges) and the presence of the trifluoromethyl side groups.

The conditions for plasma polymerization of hexafluorobenzene and tetrafluorobenzene could be adjusted in order to produce films at sufficient deposition rates with thermal stability in excess of 400°C and good adhesion to silicon and aluminum. The residual stress in these films is very high, and the dielectric constants are not as low as expected. The deposition conditions need to be improved further in order to reduce ϵ to values around 2.5. A deposition protocol has to be developed which avoids entrapment of large numbers of free radicals. The addition of hydrogen or hydrogen sources towards the end of the deposition process or the use of mixtures of argon and hydrogen containing gases (possibly hydrocarbons) for the deposition has to be checked in the future. This should result in elimination of at least a part of the entrapped radicals.

Acknowledgement

This work was supported by Siemens AG, Erlangen, Fraunhofer-Institute for Solid-State Technology (IFT), Munich, and Fonds der Chemischen Industrie, Germany.

References

1. S. P. Murarka, *Solid State Technol.* **1996 (3)**, 83
2. C. H. Ting, T. E. Seidel, *Mat. Res. Soc. Symp. Proc.* **381**, 3 (1996)
3. Semiconductor Industry Assosiation, The National Technology Roadmap for Semiconductors (1994, 1997)
4. P. Singer, *Semiconductor International* **1996 (11)**, 60
5. M. T. Bohr, *Solid State Technol.* **1996 (3)**, 105
6. S.-P. Jeng, R. H. Havemann, M.-C. Chang, *Mat. Res. Soc. Symp. Proc.* **337**, 25 (1994)
7. H. Treichel, B. Withers, G. Ruhl, P. Ansmann, R. Würl, C. Müller, M. Dietlmeier, G. Maier: „Low Dielectric Constant Materials for Interlayer Dielectrics“ in: *Low-k and High-k Materials* (H. S. Nalwa, Ed.), Academic Press, New York 1999
8. H. Treichel, G. Ruhl, P. Ansmann, R. Würl, C. Müller, M. Dietlmeier, *Microelectronic Engineering* **40**, 1 (1998)
9. D. W. Van Krevelen: *Properties of Polymers*, 3rd Ed., Elsevier, Amsterdam 1990, p. 321
10. For complete information see the MRS book series on „Low-Dielectric constant Materials“:
    Vol. I: *Mat Res. Soc. Symp. Proc.* **381** (1995), T.-M. Lu, S. P. Murarka, T.-S. Kuan, C. H. Ting, Eds.
    Vol. II: *Mat. Res. Soc. Symp. Proc.* **443** (1996), A. Lagendijk, H. Treichel, K. J. Uram, A. C. Jones, Eds.
    Vol. III: *Mat Res. Soc. Symp. Proc.* **476** (1997), C. Case, P. Kohl, T. Kikkawa, W. W. Lee, Eds.
    Vol. IV: *Mat. Res. Soc. Symp. Proc.* **511** (1998), C. Chiang, P. S. Ho, T.-M. Lu, J. T. Wetzel, Eds.
11. R. N. Johnson, A. G. Farnham, R. A. Clendinning, W. F. Hale, C. N. Merriam, *J. Polym. Sci.: Part A-1* **5**, 2375 (1967)
12. J. E. Harris, R. N. Johnson: „Polysulfones“ in: *Encyclopedia of Polymer Science and Engineering* 2nd Ed. (H. F. Mark, N. M. Bikales, C. G. Overberger, G. Menges, Eds.), John Wiley & Sons, New York 1988, Vol. 13, p. 196
13. R. May: „Polyetheretherketones“ in: *Encyclopedia of Polymer Science and Engineering* 2nd Ed. (H. F. Mark, N. M. Bikales, C. G. Overberger, G. Menges, Eds.), John Wiley & Sons, New York

1988, Vol. 12, p. 313

14. G. Maier: „Polymers with 5-membered aromatic heterocycles in the main chain" in: *The Polymeric Materials Encyclopedia: Synthesis, Properties And Applications* (J. C. Salamone, Ed.), CRC Press, Boca Raton 1966, Vol **5** H-L, p. 59
15. G. Maier, J. M. Schneider, *Macromolecules* **31**, 1798 (1998)
16. J. W. Labadie, J. L. Hedrick, *Polym. Prepr. (Am. Chem. Soc., Div. Polym. Chem.)* **31(1)**, 344 (1990)
17. J. W. Labadie, J. L. Hedrick, *Macromolecules* **23**, 5373 (1990)
18. G. Maier, R. Hecht, O. Nuyken, K. Burger, B. Helmreich, *Macromolecules* **26**, 2583 (1993)
19. K. Burger, B. Helmreich, O. Jendrewski, R. Hecht, G. Maier, O. Nuyken, *Makromol. Chem., Macromol. Symp.* **82**, 143 (1994)
20. G. Maier, R. Hecht, *Macromolecules* **28**, 7558 (1995)
21. R. Lakhan, B. Ternai: „Advances in Oxazole Chemistry“ in: *Advances in Heterocyclic Chemistry* (A. R. Katritzky, A. J. Boulton, Eds.), Academic Press, New York 1974, Vol. 17, p. 182
22. B. E. Maryanoff: „Oxazoles and Oxazolines in Organic Synthesis“ in: *Heterocyclic Compounds* (A. Weissberger, E. C. Taylor, Eds.), John Wiley & Sons, new York 1986, Vol. 45, p. 1005
23. J. A. Joule, G. F. Smith: *Heterocyclic Chemistry*, 2nd Ed., Van Nostrand Reinhold Co., London 1978, p. 315
24. R. Lakhan, B. Ternai: „Advances in Oxazole Chemistry“ in: *Advances in Heterocyclic Chemistry* (A. R. Katritzky, A. J. Boulton, Eds.), Academic Press, New York 1974, Vol. 17, p. 152
25. A. R. Katritzky, J. M. Lagowski: „Structure of Five-membered Rings with Two or More Heteroatoms“ in: *Comprehensive Heterocyclic Chemistry* (A. R. Katritzky, C. W. Rees, Ed.), Pergamon Press, Oxford 1984, Vol. 5, p. 1
26. N. Miyaura, K. Yamada, A. Suzuki, *Tetrahedron Lett.* **20**, 3437 (1979)
27. N. Miyaura, A. Suzuki, *Chem. Rev.* **95**, 2457 (1995)
28. T. Mountsier, *Proc. of the 4th DUMIC*, p. 109 (1998)

Structure and Properties of Aromatic 6-membered Polyimides Compared to 5-Membered Ones

Jin Chul Jung*, Christoph Baehr, Chang-Hun Lee, Sang-Bong Park, Young-Sub Son and Kwang S. Kim+

Polymer Research Institute/Department of Materials Science & Engineering, +Department of Chemistry, Pohang University of Science & Technology, San 31, Hyoja-dong, Pohang, 790-784 Korea

SUMMARY: When various properties of poly[2,5-di(n-alkyloxy)methyl-p-phenylenenaphthalene-1,8:4,5-diimide]s (Cm-NAPIs, m = 4, 6, 8) as aromatic 6-membered polyimides were compared with those of poly[2,5-di(n-alkyloxy)methyl-p-phenylenepyromellitimide]s (Cm-PYPIs, m = 4, 6, 8) as 5-membered ones, it was deduced that O atoms of C=O of Cm-NAPIs are not coplanar with their imide ring plane. This noncoplanarity led us necessarily to assuming that the 6-membered ring polyimide chains should have cis/trans-like isomeric structures. Detailed spectroscopic studies and ab initio calculations for appropriate model compounds are provided to support our assumption.

Introduction

Aromatic polyimides are the most important class of thermally stable polymers[1-3] currently used in many commercial applications. Of them poly(p-phenylenepyromellitimide)s are known to have the highest heat resistance[4]. This heat resistance results from their very high main chain rigidity and inter-chain attraction caused by the rigidity. However, chain rigidity is dependent not only on extent of electron delocalization along the main chain, but also on the resonance energy exerted by each group comprising the main chain. Therefore, it is readily expected that polyimide chains having larger fused aromatic rings are more rigid and heat-resistant than those having smaller fused aromatic rings, since the former possess higher hydrophobic interaction than the latter.

Out of these reasons we presumed that Cm-NAPIs having a four-fused ring system with a 6-membered imide ring should be more heat-resistant and crystalline than Cm-PYPIs having a three-fused ring system with a 5-membered imide ring. To study this problem we first prepared Cm-NAPIs and Cm-PYPIs. To the polymer backbones, we attached two (n-alkyloxy)methyl side branches to both the polyimides to enhance their solubilities[5-7], where n-alkyl groups were varied from n-butyl and n-hexyl to n-octyl. Chemical structures of these polymers are shown in Scheme 1.

 CCC 1022-1360/99/$ 17.50+.50/0

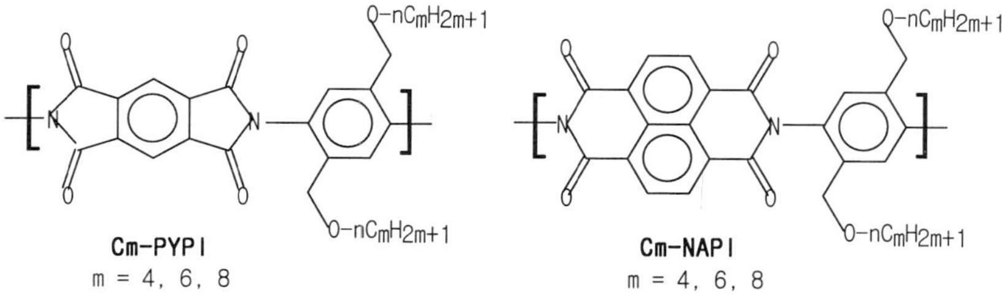

Scheme 1. Structure of the 5- and 6-membered polyimides

When physical and thermal properties of two polyimides were measured and compared to each other, we, contradictory to our expectation, found that Cm-NAPI chains are generally more flexible than Cm-PYPI chains, although a repeat unit of the former has one more ring than the latter. To elucidate this contradiction, we carried out detailed model studies by spectroscopic and ab initio-calculation methods and came to propose that the structure of a 6-membered imide ring is considerably different from that of a 5-membered one.

For these studies Cm-NAPIs and Cm-PYPIs had been prepared according to the method decribed in the literature[8,9] and N,N'-diphenyl- and dibutylnaphthalene-1,8:4,5-diimide and N,N'-diphenyl- and dibutylpyromellitimide were prepared[10-12] and used as model compounds. All the minimum energy structures and transition state structures were located with full geomety optimization by Hartree-Fock calculations using the 6-31G* basis set[13,14].

Properties

Solubility. The purpose of appending side branches to main chain was to enhance solubility of the polymers, and solubilities of Cm-NAPIs and Cm-PYPIs were checked in various solvents. All the polymers were soluble in highly acidic solvents like H_2SO_4 or CH_3SO_3H. In polar aprotic solvents Cm-PYPIs did not show any solubility at all even on heating, while C8-NAPI was soluble in N,N-dimethylpropyleneurea(DMPU) and DMAc and C6-NAPI in DMPU at room temperature. This large solubility difference indicates that NAPIs have less strong inter-chain attraction than PYPIs.

Thermal properties. Thermal resistances measured by TGA at 20°C/min heating rate under N_2 were compared between the two kinds of polyimides and their TGA thermograms are reproduced in Fig. 1. As to see from it all the PYPIs, regardless of the

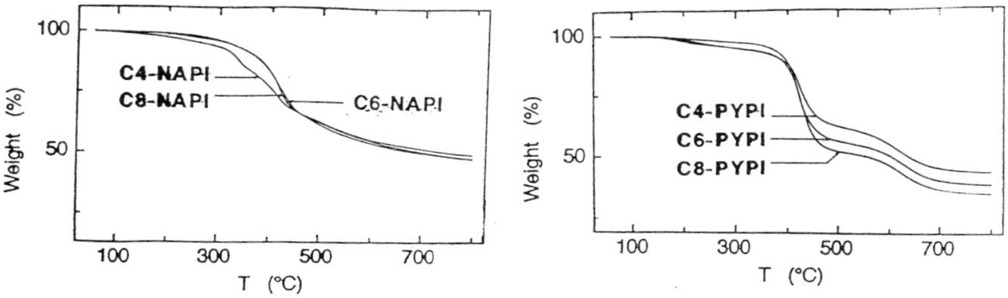

Fig. 1. TGA thermograms of Cm-NAPIs and Cm-PYPIs.

side chain length, revealed two-step pyrolysis behavior, in which side chain cleavage starts to take place at 395°C and main chain to break up at 575°C. But all the NAPIs gave smooth one-step pyrolysis behavior, in which side branches and main chain split away concurrently at 370°C. These experimental results indicate that at high temperatures there is eventually no interaction between side chain and main chain in Cm-PYPIs, while there is some in Cm-NAPIs. This means also that in interaction between main chains NAPIs are weaker than PYPIs. This weaker attraction must render NAPIs much less crystalline than PYPIs and even amorphous. The same result was really shown in WAXS study. All Cm-PYPIs are highly crystalline, while all Cm-NAPIs are completely amorphous.

UV-VIS spectra. UV-VIS spectroscopy is a powerful tool to obtain an information of extent of electron delocalization. Fig. 2 shows UV-VIS absorption behavior of the polyimides compared with the corresponding model compounds N,N'-diphenylnaphthalene-1,8:4,5-diimide(NAPI-M) and N,N'-diphenylpyromellitimide(PYPI-M). Fig. 2

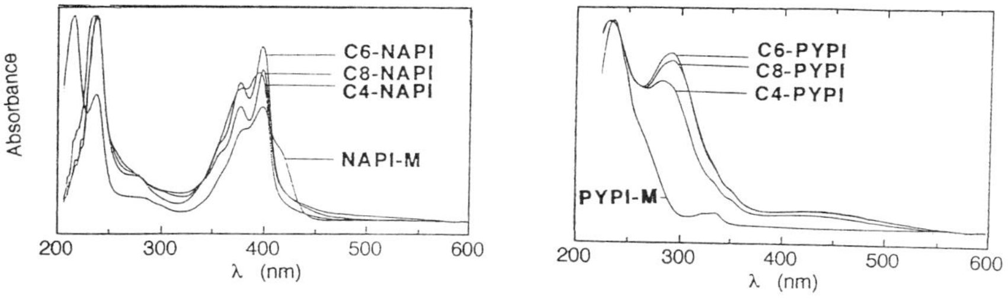

Fig. 2. UV-VIS spectra of polyimides and model compounds in H_2SO_4.

clearly shows that in the absorption maxima from both n-π* and π-π* transitions a bathochromic shift is observable between PYPIs and PYPI-M, indicating that π-electron conjugation length of PYPIs is longer than that of PYPI-M. However, between NAPIs and NAPI-M no such bathochromic shift can be observed, meaning that in NAPI main chains no π-electron delocalization is allowed. Later it was found that even the unpaired electrons on N atom can only poorly come into resonance with N-phenylene ring electrons, because N,N'-di(n-butyl)naphthaene-1,8:4,5-diimide revealed just the same UV-VIS spectrum as NAPI-M(Fig. 3). This difference in electron delocalizability along main chain between 5-membered imide ring and 6-membered one is the vital evidence of the difference in main chain rigidity between the two kinds of polyimides.

Ab initio calculation

To explain the reason why the π-electrons located on N-phenyl ring can be delocalized over 5-membered imide ring, while such delocalization is not allowed in 6-membered imide ring, ab initio calculations for N-phenylmaleimide and N-phenylnaphthalene-1,8-imide(Ph-NMI) were carried out by Hartree-Fock method using the 6-31G* basis set[13,14]. From the calculations for N-phenylphthalimide were obtained: 1) The three bonds emanating from N atom are coplanar with sp^3 configuration, 2) the phenyl ring exists 53° tilted from the imide ring plane, and 3) the energy barrier for N-phenyl ring rotation is 14kJ/mol(3.3kcal/mol). These results indicate that in N-phenylphthalimide the π-electrons of phenyl ring can be slightly delocalized over 5-membered imide ring. From calculations for Ph-NMI were obtained: 1) The three bonds emanating from N-atom are also coplanar with sp^3 configuration, 2) the phenyl tilt angle is 90° and 2) the energy barrier for phenyl ring rotation is 68kJ/mol(16.2kcal/mol). These results indicate that the π-electrons existing at N-phenyl ring can not be delocalized over 6-membered imide ring at all.

Model studies

Model compounds. The ab initio calculation results can clearly explain that π-electron delocalizability makes Cm-PYPI main chains highly rigid, while the impossible π-electron delocalization renders Cm-NAPI main chains flexible. However, these results can not explain why Cm-NAPIs are amorphous and thermally less stable than Cm-PYPIs. To answer this question we conducted model studies. First we prepared N,N'-di(n-butyl)- and N,N'-diphenylpyromellitimide(C4-BDI) and Ph-BDI) and N,N'-di(n-butyl)- and N,N'-diphenylnaphthalene-1,8:4,5-diimide(C4-NDI and Ph-NDI) from routine condensation of n-butylamine and aniline with pyromellitic dianhydride and naphthalene-1,8:4,5-tetracarboxylic dianhydride, respectively.

UV-VIS spectra. For the model compounds we measured UV-VIS spectra again in concentrated H_2SO_4. These spectra are reproduced in Fig. 3. It clearly shows that Ph-BDI's maximum exists at lower wave length than C4-BDI, while Ph-NDI's spectrum is totally the same as C4-NDI's, although the N-substituent is changed from aromatic phenyl to aliphatic n-butyl. This result confirms again that no π-electrons located on the N-phenyl ring can come into conjugation over 6-membered imide ring, while it is allowed in 5-membered one.

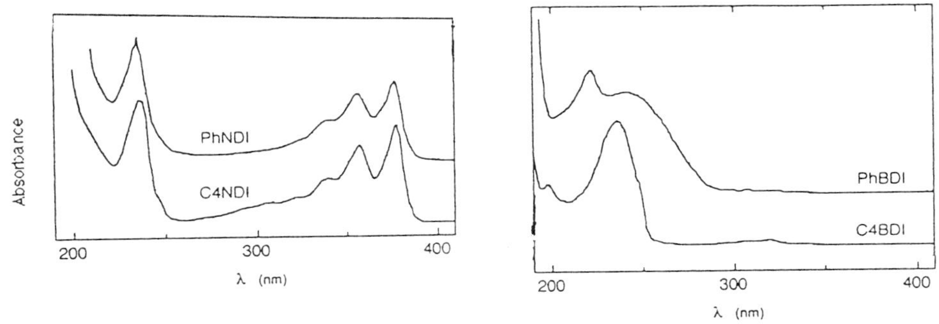

Fig. 3. UV-VIS spectra of the model compounds in H_2SO_4.

Solid-state ^{13}C-NMR spectra. The ab initio calculation has shown that the N-phenyl ring of N-phenylphthalimide is 53° tilted from 5-membered imide ring plane, while that of N-phenylnaphthalene-1,8-imide is 90° tilted from 6-membered imide ring plane. Based on this result it is reasonable to assume that two N-phenyl rings of Ph-BDI and Ph-NDI are also 53° and 90° tilted, respectively. If this assumption is true, we must naturally come to the deduction that in solid state the two N-phenyl rings of Ph-BDI must be in either parallel(53°/53°) or antiparallel(53°/147°) conformation and the two N-phenyl rings of Ph-NDI must have only one conformation(90°/90°). To confirm this deduction solid-state ^{13}C-NMR spectra of Ph-BDI and Ph-NDI were taken. Our interpretation of the ^{13}C-NMR spectrum of Ph-BDI could not find out any presence of two conformations, meaning that the above deduction is not true. However, in the spectrum of Ph-NDI a clear doublet of C=O carbon absorption was found, as to see from Fig. 4. This result means that in solid state Ph-BDI has only one conformation, while Ph-NDI has two conformations giving rise to two C=O carbons which are in different chemical environments.

To get more detailed understanding on the reason of C=O splitting into a doublet in Ph-

NDI, we also took a solid-state ^{13}C-NMR spectrum of C4-NDI. This spectrum exhibited only a singlet C=O carbon, meaning that the four C=O carbons of C4-NDI are chemically equivalent, and hence, C4-NDI has only one conformation in solid state. This spectral difference between C4-NDI and Ph-NDI results naturally from replacement of n-butyl group by phenyl group, meaning that the C=O absorption splitting of Ph-NDI arises from a certain influence of phenyl ring.

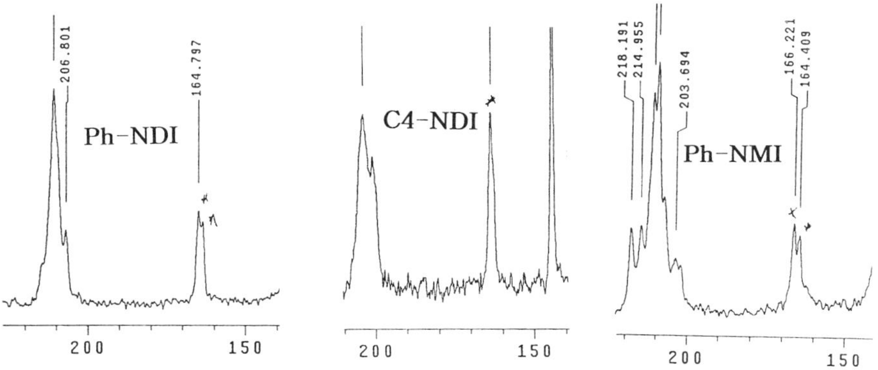

Fig. 4. MAS-^{13}C-NMR spectrum of Ph-NDI, C4-NDI and Ph-NMI.

To explain the presence of these two conformations, we made a simple molecular modeling calculation to obtain energy minimized configuration and torsional energy profile of N-phenylnaphthalene-1,8-imide(Ph-NMI) by CS Chem 3D program (Cambridge Scientific Computing, Inc.). The numerical torsional energy values obtained by rotating 180° around the N-C_{ph} were calibrated so that the maximum value corresponds to the value 16.2kcal/mol(68kJ/mol) obtained from the ab initio calculation, because absolute energy values calculated by simple simulation programs are not highly reliable[16)].

Torsional energy profile thus obtained is reproduced in Fig. 5. It shows that in the most stable state the N-phenyl ring is tilted about 45° from 6-membered imide ring plane. The destabilized state at 90° is the state formed by complete loss of resonance energy of 3.5kcal/mol. Fig. 5 shows further that there are two stable states, one at 45° and the other at 135°, but the 135°-state is a little bit more stable than the 45°-state. The result that even Ph-NMI which half-models the NAPI polymer has two different states must indicate that the two C=O groups contained in it must also be different from each other. This difference should be observable in ^{13}C-NMR spectroscopy. In Fig. 4 its solid-state

MAS-^{13}C-NMR spectrum is included. As it shows, this half-model compound has really two C=O peaks, which confirms the presence of two different C=O groups in it.

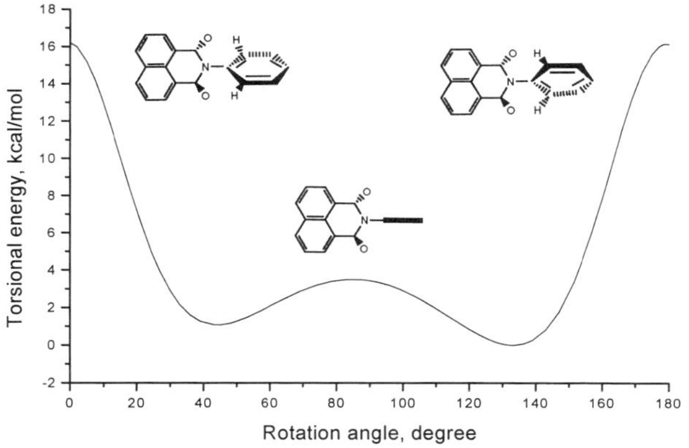

Fig. 5. Torsional energy profile of Ph-NMI around N-C_{ph} bond

To explain these two different C=O groups, we assumed that the O atoms of C=O groups are not coplanar with the 6-membered imide ring plane, but a little bit distorted. When the imide plane is laid on paper plane, one O atom might direct downward and the other upward from paper plane. This distortion is caused by the hindered steric interaction between two H atoms at ortho position of N-phenyl ring and two O atoms bonded to C=O groups of 6-membered imide ring. This steric hindrance leads to a difficulty of free rotation of phenyl ring around N-C_{Ph} bond, resulting to an atopoisomerism. This steric hindrance is not exerted in 5-membered imide ring, because the H-O distance is sufficiently long. This difference in rotation difficulty between 5- and 6-membered imide ring has already been evidenced by the result that torsional energy of 5-membered ring is only 3.3kcal/mol, while that of 6-membered one is 16.2kcal/mol. 3.3kcal/mol is certainly too small to observe in NMR.

Fig. 5 shows that Ph-NMI has resonance energy of 3.5kcal/mol. However, this value is apparently too small to observe in UV-VIS spectroscopy in solution, because in solution this small energy barrier can readily be overcome by rapid molecular movements.

As already described above, the 135°-state is a little bit more stable than the 45°-state. This stability difference can be readily understood when we consider the distance between $O_{C=O}$ and H atom at ortho position of phenyl($H_{O\text{-}Ph}$) in both the states. Two $H_{O\text{-}Ph}$

atoms come farther from $O_{C=O}$ in 135°-tilted state than in 45°-tilted state, as depicted in Fig. 5. Although this stability difference is quite small, and it can not be observable in solution. However, it is large enough to observe from solid-state ^{13}C-NMR spectroscopy of Ph-NMI, as shown in Fig. 4.

The fact that both Ph-NMI and Ph-NDI have a doublet C=O absorption in solid-state ^{13}NMR spectroscopy means that the C=O splitting of one 6-membered imide ring is not affected by the presence of the other one, although they are bound at the same molecule across the naphthalene ring. If it were affected, Ph-NDI should reveal more number of C=O peaks than a doublet.

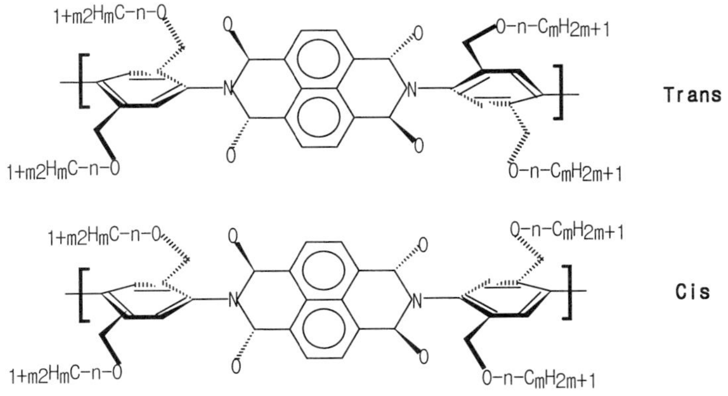

Scheme 2. Proposed chain structure of Cm-NAPIs

When we now consider chain structure of NAPIs in solid state on the basis of our result of the isomeric structures of Ph-NMI and Ph-NDI, each N-phenylimide group can be in two states, regardless whether a naphthalene ring is substituted by one imide group or by two. This means that each NAPI repeat unit has two stable states in solid state. When these two states are depicted along the main chain of NAPI, there must be cis/trans-like structures, depending on orientation of C=O and phenyl groups. These cis and trans structures are depicted in Scheme 2. These cis/trans-like structures to be present in Cm-NAPIs can explain that they are amorphous in solid.

Conclusions

Unlike a 5-membered imide ring an N-aromatics-substituted 6-membered imide ring has two isomeric states in solid state, depending on orientation of C=O and N-aromatic group. This isomerism arises from the slightly hindered rotation of N-aromatic group. It leads to presence of cis/trans-like isomeric chain structures in 6-membered polyimides. The presence of these isomeric chain structures result to more flexible main chain nature, more amorphous morphology, lower thermal resistance and higher solubilities in organic solvents than the corresponding 5-membered polyimides.

Acknowledgement. The authors thank the Korea Science & Engineering Foundation (95-0501-08-3), the Deutsche Forschungsgemeinschaft and the Center for Advanced Functional Polymers for their financial supports.

References

[1] C. E. Sroog, *J. Macromol. Sci., Macromol. Rev.*, **11**, 161 (1976)

[2] J. C. Jung, "*Processable Polyimides*", in J. Salamone Ed., "*Polymeric Materials Encyclop.*", CRC Press, New York London Tokyo, **8**, 6292 (1996)

[3] H. Kim, Y.-J. Choi, J. C. Jung, W.-C. Zin, *Polym. Bull.*, **38**, 689 (1997)

[4] M. I. Bessenov, M. M. Koton, Kudryavtsev, L. A. Laius, "*Polyimides*", Consultants Bureau, New York & London 1987

[5] M. Ballauff, *Angew. Chem. Int. Ed. Engl.*, **28**, 253 (1989)

[6] M. Wenzel, M. Ballauff, G. Wegner, *Makromol. Chem.*, **188**, 2865 (1987)

[7] F. Helmer-Metzmann, M. Rehahn, L. Schmitz, M. Ballauff, G. Wegner, *Makromol. Chem.*, **193**, 1847 (1992)

[8] J. C. Jung, S.-B. Park, *Polym. Bull.*, **35**, 423 (1995)

[9] J. C. Jung, S.-B. Park, *J. Polym. Sci., Polym. Chem. Ed.*, **34**, 357 (1996)

[10] R. A. Meyers, *J. Polym. Sci., Part A-1*, **7**, 2757 (1969)

[11] N. D. Ghatge, U. P. Mulik, *J. Polym. Sci., Polym. Chem. Ed.*, **18**, 1905 (1980)

[12] P. S. Carleton, W. J. Farrissey, Jr., J. S. Rose, *J. Appl. Polym. Sci.*, **16**, 2983 (1972)

[13] K. S. Kim, B. J. Mhin, U.-S. Choi, K.-J. Lee, *J. Chem. Phys.*, **97**, 6649 (1992)

[14] C. Cui, S. J. Cho, K. S. Kim, C. Baehr, J. C. Jung, *J. Chem., Phys.*, **107**, 10201 (1997)

[15] F. Helmer-Metzmann, M. Ballauff, R. C. Schulz, G. Wegner, *Makromol. Chem.*, **190**, 985 (1989)

[16] H. R. Kricheldorf, V. Linzer, *J.M.S.-Pure Appl. Chem.*, **A32**, 311 (1995)

Photocrosslinking of Thin Polymer Films – Materials for Sensors and Actuators

Dirk Kuckling[1] *, Hans-Jürgen Adler[1], Karl-Friedrich Arndt[2], Thomas Wolff[2], Jan Hoffmann[3], Wolf-Joachim Fischer[3]

[1] Institute of Macromolecular Chemistry and Textile Chemistry,
[2] Institute of Physical Chemistry and Electrochemistry,
[3] Institute of Semiconductor and Microsystem Technology,
Dresden University of Technology, D-01062 Dresden, Germany

SUMMARY: Polymers with temperature dependent degrees of swelling, especially polymers which exhibit lower critical solution temperature (LCST) behaviour in aqueous solutions, are of interest for applications in microsystems. For these applications it is necessary to form and pattern thin films in the μm-range. This has been achieved through photocrosslinking of linear prepolymers.
Copolymers based on N-isopropylacrylamide (NIPAAm) were modified with a stilbazolium salt chromophore to yield photocrosslinkable temperature sensitive polymers. The chromophore reacts via a [2+2]-cycloaddition under irradiation, this can be used to crosslink the polymer. The photocrosslinking properties were studied by UV irradiation of thin films and measuring the changes in UV absorption spectra. Through irradiation of thin films through a mask it was possible to obtain patterned networks in the μm-range (20 μm space width and > 50 μm line width). The polymers exhibited LCST behaviour, which was measured using DSC. The resulting patterned networks had temperature dependent swelling properties in aqueous media.

 CCC 1022-1360/99/$ 17.50+.50/0

Introduction

The properties of polymer gels are between those of a liquid and a solid. These properties make them very interesting for intensive studies. Polymer gels can be easily deformed by external stimuli, so that they generate force or execute work on the external environment. If such responses can be translated from the microscopic level (i. e. dimensions of polymer chains) to a macroscopic scale, the conversion of chemical free energy into mechanical work should be realized. As early as 1948 it was found that water-swellable polymer gels can convert chemical energy directly into mechanical work.[1,2] Since then such systems were called "Chemomechanical systems". These first studies were carried out on polyelectrolyte gels changing their state of ionization.[3] Later phase transition behaviour was also found to occur with nonionized gels.[4] The main focus in this direction is the investigation of polymers which exhibit a lower critical solution temperature behaviour (LCST). A famous example of this class is poly(N-isopropylacrylamide) (PNIPAAm), which shows a reversible phase transition coupled with swelling-deswelling upon heating and cooling.[5-9] However, there are other mechanisms e. g. photoinduced transitions.[10,11]
To use these effects in actuators the change of the swelling degree must also occur under pressure. To maximize the executed work the changes in swelling degree and the swelling pressure must be optimized. Therefore, phase transitions in polymer gels are of great interest, because they are often accompanied with a big change in swelling degree and during phase transition with a strong swelling pressure. The most important point is that all this is often caused by a small change of the environmental conditions.

Results and Discussion

Bulkgels (i. e. gel samples in the mm-range) can easily be prepared and modified and the effects of changes in the environmental parameters can be investigated. But the kinetics of swelling and deswelling in these gels are typically governed by diffusion-limited transport of the polymeric components of the network in water, the rate of which is inversely proportional to the square of the smallest dimension of the gel.[6,12] Thus, the response times of the gels are very slow. However, it is possible to decrease the response time, e. g. by changing the synthesis conditions.[13]

Since the kinetics of swelling and deswelling are proportional to the gel dimensions, reduction of the gel size to the µm-scale should be very effective in decreasing the response time.[14] In such cases applications in microsystems are conceivable.[15] In order to investigate these systems networks of sensitive polymers in the µm-range were prepared.

As a result of this work it could be shown that photocrosslinking of linear polymers based on PNIPAAm is a potential way to solve this problem.[16] Photocrosslinking of a linear polymer is advantageous of two reasons: Thin films can be easily prepared and networks can be formed from defined polymers. As a photocrosslinking reaction, a well defined reaction like a [2+2]-cycloaddition should be used to control the crosslinking process. A stilbazolium salt was chosen as the chromophore, because it consists of two structural parts. One is a hydrophobic moiety (stilben backbone), which is necessary for the crosslinking reaction and the other is a hydrophilic domain (pyridinium group with a positive charge), which was selected to compensate for the decrease in the LCST caused by the hydrophobic substituents. The chromophore can be synthezised in three steps from common chemicals.[16]

Several PNIPAAm copolymers with acrylamides bearing carboxylic groups were prepared by free radical polymerization.[17] Acrylic acid can also be used instead of this special monomer. The polymer analogous reaction of these polymers with the chromophore yields PNIPAAm-photopolymers containing different amounts of chromophore.

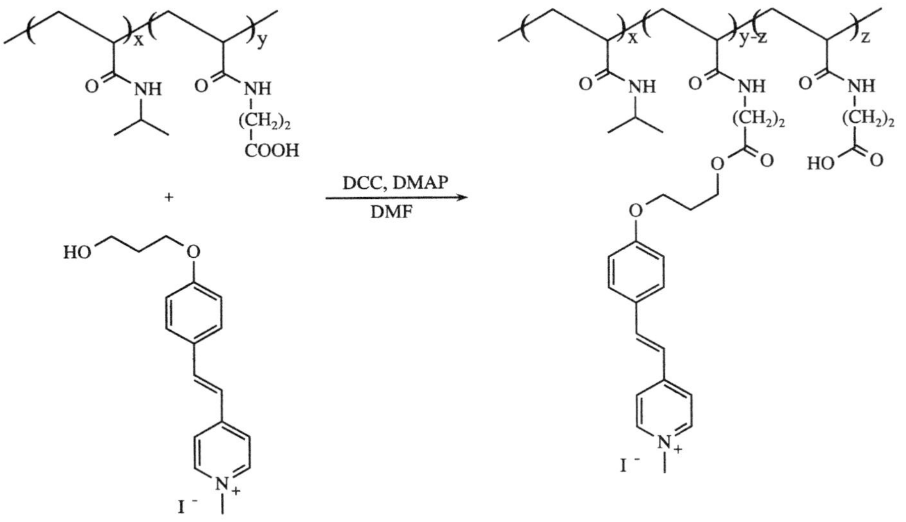

Fig. 1: Preparation of the photocrosslinkable polymers

In Table 1 the characteristics of the photocrosslinkable polymers are shown. The molecular weights are between 40 000 and 60 000 g/mol. By varying the amount of comonomer in the polymer different amounts of the chromophore could be introduced, ranging from a very low content up to a very high content. The cloud point of the prepolymer, measured by DSC, was not influenced by the comonomer content, remaining in the range of homo-PNIPAAm. But after modification the cloud point was strongly effected by the chromophore. Increasing the chromophore content decreased the cloud point of the photopolymer. Finally at very high chromophore contents the LCST behaviour disappeared.

Tab. 1: Characterization of the photopolymers

no.	polymer composition ($x_{comonomer}$)	M_w [g/mol]*	chromophore content	cloud point**	
				prepolymer	photopolymer
I	0.025	59 000	0.2 %	31.3 °C	25.6 °C
II			1.2 %		24.3 °C
III	0.053	57 000	3.2 %	31.9 °C	17.6 °C
IV	0.100	40 000	7.0 %	31.9 °C	14.1 °C
V			12.3 %		-

* with GPC (THF/0.3 % NMAA)

** with DSC (5 wt.-% solution, heating rate 5 °C/min)

As has been mentioned, two chromophore molecules are able to react via a [2+2]-cycloaddition. Coupling the chromophore with a polymer chain followed by irradiation with light of the correct wavelength leads to the formation of networks.

The formation of such junction points can be seen by a decrease of the UV absorption of the chromophore.[18] Thin films were prepared on quarz plates and irradiated with light. The UV absorption was measured after different irradiation times. Fig. 2 shows the decrease of the UV absorption at ca. 380 nm with time. This indicates that a [2+2]-cycloaddition of the stilbazolium moieties occurs, which destroys conjugation in the entire π-electron systems resulting in a decrease in the UV absorption of the stilbazolium unit. By inreasing the irradiation time to longer than 600 min the absorption could be further reduced.

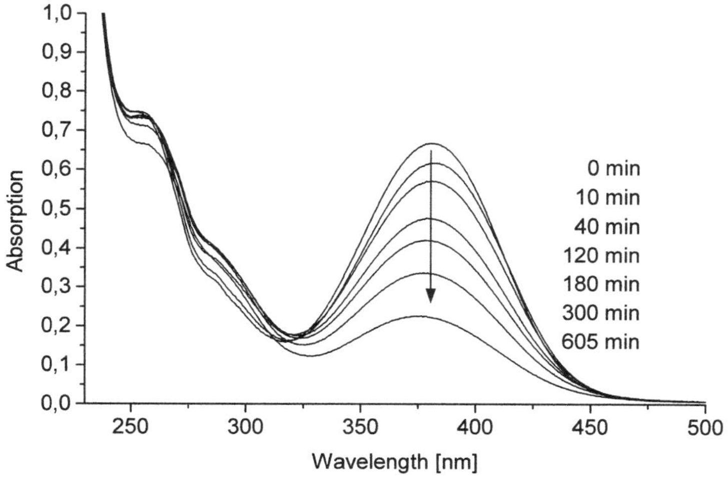

Fig. 2: Changes in the UV spectra of the photopolymer III during irradiation

In Fig. 3 the irradiation kinetics are shown. The crosslinking efficiency increased with the concentration of the chromophore. At very high chromophore contents (sample with 12.3 %) an irradiation time of only 5 min was necessary to form an unsoluble network. The stilbazolium units associate in solution and the degree of association depends on the stilbazolium concentration. After film preparation these associates were fixed in the dry film. A higher amount of associates led to a higher crosslinking efficiency. On the other hand a chromophore content of only 0.2 % was apparently not enough to form a network even with high conversion of the chromophore. Further investigations were focused on polymers with a medium content of chromophore.

One main aim of this work is the preparation of sensitive networks in the µm-range. Therefore, photocrosslinkable sensitive polymers were synthesized in order to use them in a micropatterning process, well-known in the microsystem technology. The preparation steps were as follows:[15] A thin polymer layer was produced by spin coating using polymer solutions. The film thickness in the dry state depends on a variety of parameters such as the concentration of the solution, the rotation speed during the coating process etc. The film thickness can be adjusted to between 0.5 µm and 30 µm. The next step was the photocrosslinking through a photomask. The network can only be formed in the irradiated areas. The final step was the development of the pattern by dissolving the uncrosslinked polymers. In this case water was found to be a suitable solvent.

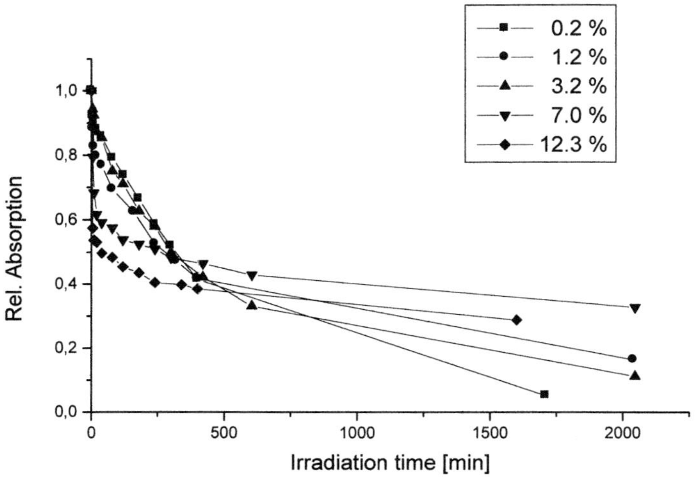

Fig. 3: Irradiation kinetics of the photopolymers

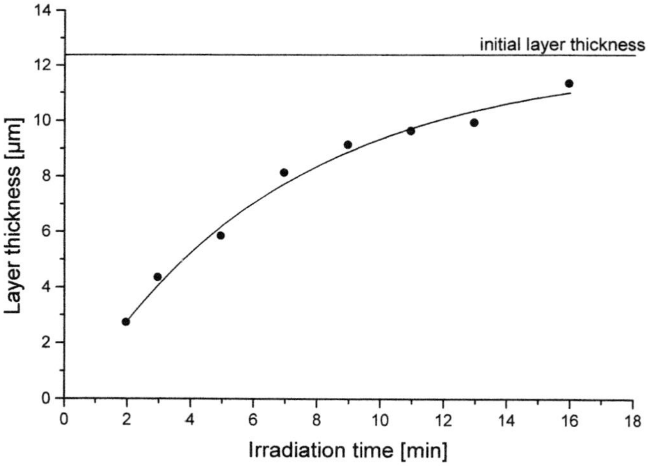

Fig. 4: Growth of layer thickness during irradiation

Another important influence on the layer thickness of the pattern was the irradiation time. In Fig. 4 it could be seen that an increasing irradiation time increased the layer thickness. During irradiation more junction points were formed and more polymer was fixed, resulting in an increase in the film thickness. The film thickness exponentially approached the value of the initial film thickness. With an increase of the irradiation time not only the amount of fixed polymer increased but also the crosslinking density. The crosslinking density plays an important role in the resolution properties of the patterning process. A loosely crosslinked network can swell during the development to very high degrees of swelling. This resulted in a high stress on the adhesion of the network to the wafer, which normally caused the network to lose contact with the wafer. The result can be seen in Fig. 5a (after 3 min irradiation time). An increase in irradiation time increased the crosslinking density and less defects in the pattern were formed. After 7 min and 11 min, respectively, the patterns are well formed for such highly swellable polymer layers.

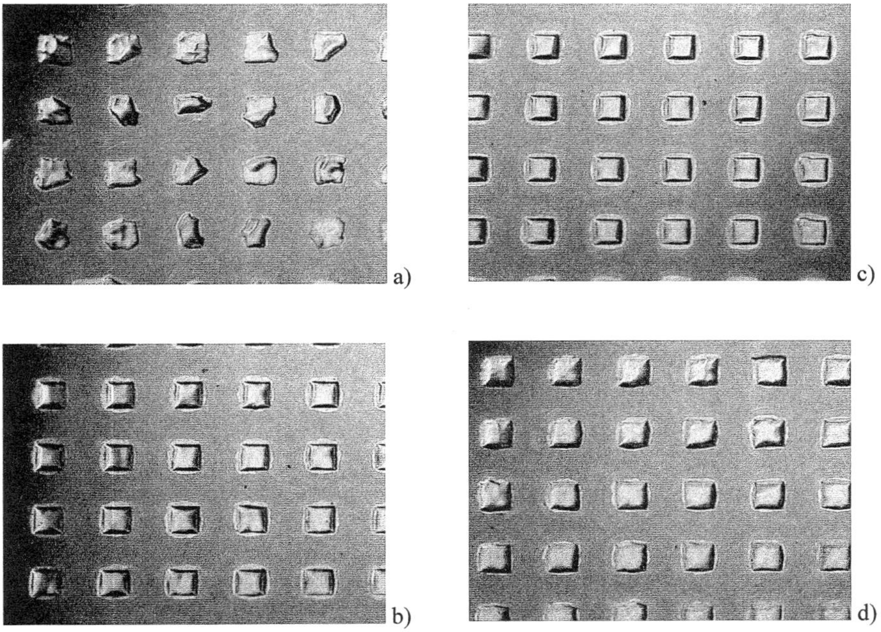

Fig. 5: Pattern of PNIPAAm-Gels after different irradiation times (3 min, 5 min, 7 min and 11 min, respectively)

To test the resolution properties of the patterning process a lot of patterns were created. One example is shown in Fig. 6. The complete wheel structure (Fig. 6a) was well depicted. The view of the outer part of this structure (Fig. 6c) shows that this structure consists of long fine polymer stripes. The view of the inner part of the structure (Fig. 6b) shows that the width of the stripes is decreased. As a result the contact area of the network on the wafer is decreased. The polymer lost contact during development and some worm like structures were obtained.

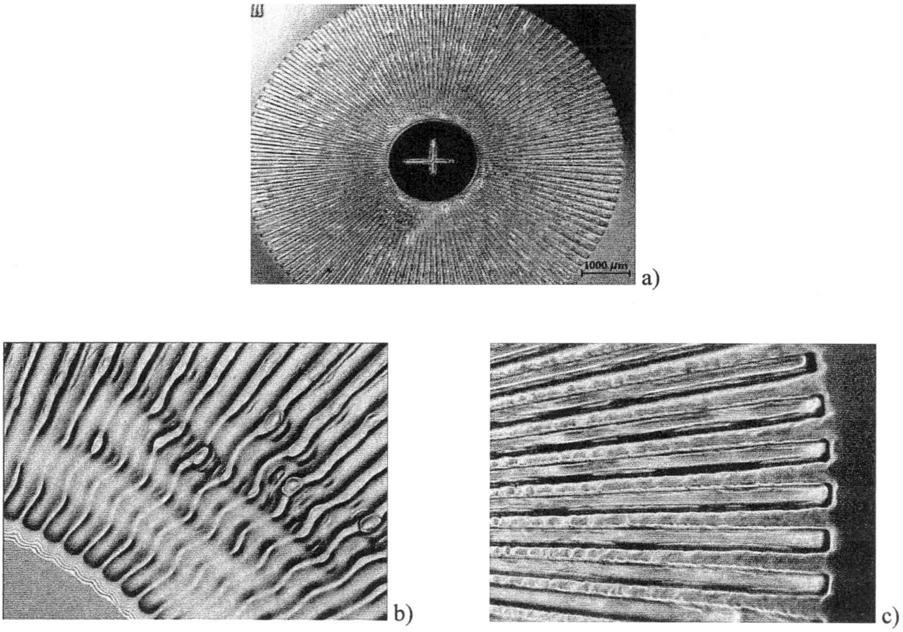

Fig. 6: Resolution properties of the patterning process

It was possible to demonstrate the temperature sensitivity of the networks. In order to do this the square patterns were immersed in water and the gel thickness could be measured by viewing the side of the pattern with a microscope. Fig. 7 shows the temperature dependent degree of swelling. At low temperatures the gel was in the swollen state. Different degrees of swelling resulted from different irradiation times and thus from different degrees of crosslinking. The gel began to shrink when the temperature was raised to above the LCST (and this is approximately the same as in the uncrosslinked state). At temperatures above 25 °C the gel was in the shrunken state independent of the degree of crosslinking. The properties of PNIPAAm bulkgels could be transfered into thin layers. The gels reached the equilibrium degree of swelling within a few seconds.

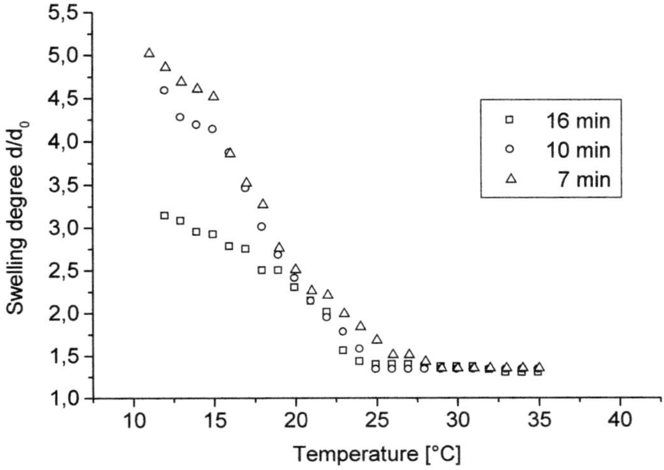

Fig. 7: Change in degree of swelling of thin crosslinked PNIPAAm-layers

Conclusion

Though the combination of temperature sensitive polymers and photodimerizable chromophores it is possible to obtain temperature sensitive networks in the μm-range. The reduction of the gel dimensions also decreases the response time of the gels down to seconds and these networks should be useful for actuators and sensors in microsystems.

Acknowledgement

The DFG (Deutsche Forschungsgemeinschaft) are gratefully acknowledged for their financial support of this work within the "Sonderforschungsbereich 287". The authors would like to thank W. D. Habicher (Institute of Organic Chemistry, TU Dresden) for his cooperation.

References

[1] W. Kuhn, *Experimentia* **5**, 318 (1949)

[2] W. Kuhn, B. Hargitay, A. Katchalsky, H. Eisenberg, *Nature* **165**, 515 (1950)

[3] S. Hirotsu, Y. Hirokawa, T. Tanaka, *J. Chem. Phys.* **87**, 1392 (1987)

[4] Y. Hirokawa, T. Tanaka, *J. Chem. Phys.* **81**, 6379 (1984)

[5] A. S. Hoffman, *Macromol. Symp,* **98**, 645 (1995)

[6] S. H. Gehrke, *Adv. Polym. Sci,* **110**, 81 (1993)

[7] M. Shibayama, T. Tanaka, *Adv. Polym. Sci.* **109**, 1 (1993)

[8] S. Hirotsu, *Adv. Polym. Sci.,* **110**, 1 (1993)

[9] H. G. Schild, *Prog. Polym. Sci.* **17**, 163 (1992)

[10] D. Kungwatchakun, M. Irie, *Makromol. Chem., Rapid Commun.* **9**, 243 (1988)

[11] H. Menzel, R. Kröger, M. L. Hallensleben, *Macromol. Report,* **A32**, 779 (1995)

[12] T. Tanaka, D. J. Fillmore, *J. Chem. Phys.* **70**, 1214 (1979)

[13] B. G. Kabra, S. H. Gehrke, *Polymer Commun.* **32**, 322 (1991)

[14] S. Zhou and C. Wu, *Macromolecules* **29**, 4998 (1996)

[15] J. Hoffmann, M. Plötner, D. Kuckling and W.-J. Fischer, *Sens. Actuators*, accepted

[16] D. Kuckling, H.-J. P. Adler, K.-F. Arndt, J. Hoffmann, M. Plötner, T. Wolff, *Polym. Adv. Technol.*, accepted

[17] D. Kuckling, H.-J. Adler, K.-F. Arndt, L. Ling and W. D. Habicher, *Macromol. Chem. Phys.*, submitted

[18] R. Schinner, T. Wolff, D. Kuckling, *Ber. Bunsenges. Phys. Chem.* **102**, 1710 (1998)

Self-Assembled Multilayers of Photochromic Bola-amphiphile and Ionene-type Oligomer: Relation of Aggregate State and Photoisomerization

Ae-Li Park and Jong-Dal Hong*

Dept. of Chemistry, Inchon University, 177 Dohwa-dong Nam-ku, 402-749 Inchon (sKorea)

SUMMARY: Photoisomerization process of azobenzene has been studied in the multilayer assemblies of poly(vinylsulfate, potassium salt) and two different layer-elements; 4,4'-azobenzene-bis(12-trimethylammonium-N-yl dodecyloxy) dibromide (BA-12, **1**) and an ionene oligomer, poly(4,4'-azobenzene-bis(12-dimethyl-ammonium-N-yl dodecyloxy) bromide) (PAZ-12, **2**). The change of the photoactive layers from BA-12 to PAZ-12 was found to dramatically influence the photo-isomerization characteristics of the azobenzene in the multilayer. Surprisingly, we have observed that the chromophore interaction in PAZ-12 is stronger than that in BA-12 in self-assembled (SA) film (λ_{max} = 320 nm compared to 341 nm), although the aggregate formation is considered to be hindered by the strains of the polymer chains. For BA-12/PVS and PAZ-12/PVS multilayers, the steady state between cis and trans isomers reached after ca 20 min. and 48 min. of UV irradiation, respectively. It was also found that after first one trans-cis-trans cycle of the photoisomerization, the trans to cis isomerization was completed in 30 sec of irradiation in the following 10 cycles without appearance of any fatigue in the multilayer film.

Introduction

During past several years, the layer-by-layer self-assembly technique based on electrostatic attraction between opposite charges[1] has been devoted to realization of the variety of the multilayer heterostructures on a solid substrate. The assembly of cationic and anionic polyelectrolytes[2] has been extended to the polyelectrolytes and DNA[3], polyelectrolytes and latex particles[4], polyelectrolytes and proteins[5], polyelectrolytes and delaminated clay platelets[6-9], polyelectrolytes and colloidal metal particles[10-12], polyelectrolytes and dyes[13-18], polyelectrolytes and electrically conductive or the electroluminescent conjugated polymers[19]. Bolaamphiphiles have been used as components of this technique[1, 20]. The layer formation by these bipolar molecules produces charged surfaces upon their adsorption and allows the alternate assembly to continue. As can be also seen from the precedents of bolaamphiphiles and bilayer-forming amphiphiles, small molecules that tend to aggregate in water have been assembled as molecular layers in combination with oppositely-charged polyions. Very recently, photoisomerization was demonstrated in the multilayer assemblies, in which the photoactive

 CCC 1022-1360/99/$ 17.50+.50/0

Scheme I

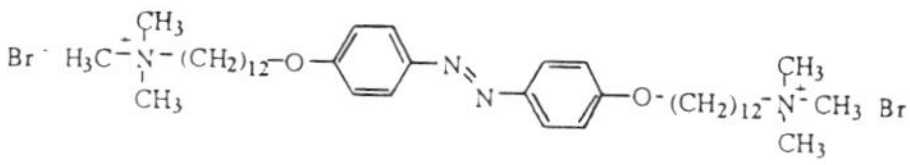

BA-12, **1**

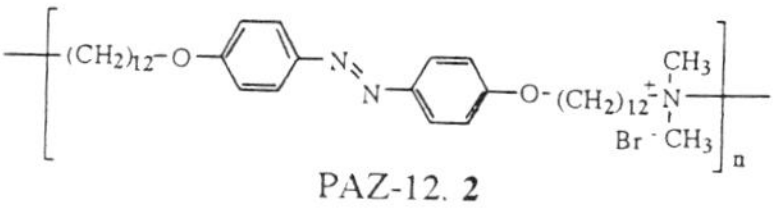

PAZ-12, **2**

layers sandwiched between oppositely-charged polyelectrolyte layers consist of bolaamphiphiles bearing azobenzene chromophores[21-23].

The reversible cis-trans photoisomerization is of great interest for application in optical information storage systems. The azobenzene molecules undergo "trans-cis" reversible photoisomerization under irradiation of photoactive light with an appropriate wavelength. These changes of the molecular structure result in modified intermolecular interactions and in a modification of the supramolecular order of the SA multilayers. So, the periodicity of the lamellar architecture and the in-plane order is affected. Laschewsky and his coworkers[24] reported about thin polymeric, non-centrosymmetrical coatings of polyion containing azobenzene moiety based on layer-by-layer electrostatic SA methods, but studies on the photoisomerization of azobenzene in SA polymeric multilayer films are still lacking.

In order to study the photochromic process, we prepared a water-soluble bolaamphiphile, 4,4'-azobenzene-bis(12-trimethylammonium-N-yl dodecyloxy) dibromide (BA-12, **1**) and an ionene oligomer, poly(4,4'-azobenzene-bis(12-dimethyl-ammonium-N-yl dodecyloxy) bromide) (PAZ-12, **2**) bearing a photochromic chromophore, as shown in Scheme I. Additional interest arises from the question as to whether and to what extent the photoisomerization will take place in the multilayer assemblies, in which the molecular mobility in the photochromic layers is restricted by the oppositely-charged polyelectrolyte layers.

The aim of this article is to study the influence of the supramolecular order on the photoisomerization characteristics of azobenzene in the multilayer assemblies by varying the component of the photoactive layers sandwiched between oppositely-charged polyelectrolytes layers from the small molecule, bolaamphiphile BA-12 to the ionene oligomer PAZ-12.

Materials and Methods

Materials. Poly(vinylsulfate, potassium salt) (MW = 140,000) PVS and poly(allylamine hydrochloride) (Mn = 50,000 ~ 65,000) PAH were purchased from Aldrich and used without further purification. The synthesis and characterization of the bolaamphiphile BA-12 were described in the literature[23]. The ionene PAZ-12 was prepared by an addition polymerization of the corresponding diamine and dibromide, which were synthesized according to the literature process[25]. The detailed description of the synthesis and characterization of PAZ-12 will be published in a forth-coming paper.

Film Construction. The ultrapure water used for all experiments and all cleaning steps was obtained by ion-exchange and filtration unit (Milli-Q, Millipore GmbH). The resistivity was better than 18.0 MΩ · cm. The substrates for all adsorption experiments were fused silica slides of size 25x50 mm^2. They were cleaned by ultrasonication in the mixture of H_2SO_4/H_2O_2 (7/3) and then heated in the mixture of $H_2O/H_2O_2/NH_3$ (5:1:1) at 80°C for 1 hr. The substrates were thoroughly washed with ultrapure water after both steps.

The bolaamphiphile, ionene oligomer and polyelectrolytes were deposited onto the negatively-charged substrate as described previously[1-2)]. After each adsorption step, the surface of the film was thoroughly rinsed and then blown dry with a stream of nitrogen. In order to adsorb BA-12 or PAZ-12, the substrate was precoated with five bilayers of PAH and PVS for providing a more homogeneous charge distribution on the substrate.

Multilayer film was achieved by dipping the substrate alternately into the aqueous solution of BA-12 (0.2 mM), and the aqueous solution of PVS (6 mM). Thus, the substrate was coated with ten bilayers of BA-12 and PVS.

Multilayer film of PAZ-12 and PVS was deposited by dipping the substrate alternately into the DMSO/H_2O (10/3) solution of PAZ-12 (0.5 mM) and the aqueous solution of PVS (6 mM). The quantity of material deposited at each step was determined from its absorption spectrum, which was generated on a UV/Visible spectrophotometer (Perkin-Elmer Lambda 40).

Photoisomerization. Photoisomerization of the azobenzene was induced by UV light (360 nm) for the trans-cis reaction and by visible light (blue, 450 nm) for the cis-trans back isomerization. The irradiations were carried out using a high-pressure mercury lamp (Altech, 1 kW) equipped with glass filters (ultraviolet: UG1, Spindler & Hoyer, blue: GG10, Spindler & Hoyer). The lamp power was adjusted to 2 mW/cm^2 for UV light and to 100 mW/cm^2 for blue light. The surrounding temperature on the sample was adjusted to be ca 30°C using a cold plate (Stir-Kool Model SK12, Aldrich Chem.Co.) to prevent thermal back isomerization during UV irradiation.

Film Morphology. The film thickness of multilayer films on a silicon wafer was determined with an optical ellipsometer (Rudolph/Auto EL) equipped with a 632.8 nm line of He/Ne laser incident upon the sample at 70°. The orientation of the azobenzene chromophore in BA-12/PVS and PAZ-12/PVS multilayer was estimated from the polarized UV spectra obtained at 45° incidence angle.

Results and Discussion

1. Alternating Multilayer Assemblies of Bolaamphiphile and Polyelectrolytes

1.1. Preparation of Multilayer Assemblies

In order to study the photoisomerization characteristics of the multilayer assemblies in which the photoactive layers are sandwiched between two polyelectrolyte layers, we have constructed a SA multilayer film composed of a cationic bolaamphiphile BA-12 and an anionic

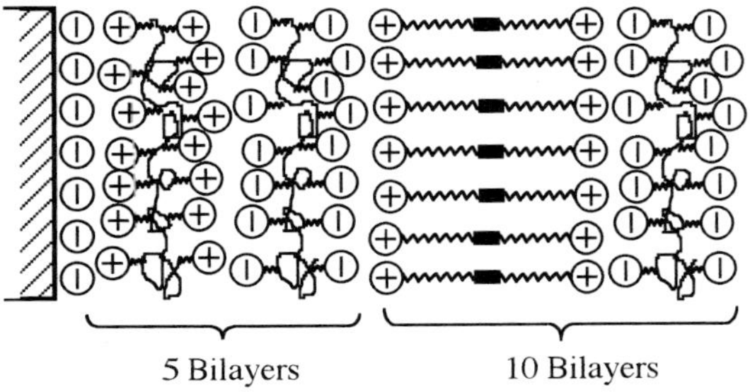

Fig. 1: Idealized schematic structure of the multilayer assemblies composed of the BA-12 dication and anionic polyelectrolytes PVS.

polyelectrolyte PVS on a fused silica precoated with five bilayers of PAH and PVS, as shown schematically in Fig. 1. Photoactive BA-12 layer exists between two polyelectrolyte layers in order to allow for a photoisomerization to take place.

Fig. 2 shows the UV absorbance of azobenzene chromophore at 341 nm as a function of the number of bilayers deposited. The absorption occurs essentially uniformly, the same amount being deposited on each cycle. A linear fit of data yields an average increase of the optical density of 0.018 per layer of BA-12. Additionally, the absorbance of BA-12 in the multilayer assemblies differs essentially not from that in the aqueous solution. This suggests that the aggregates formed in solution were directly transferred on the substrate without big change of the aggregate state.

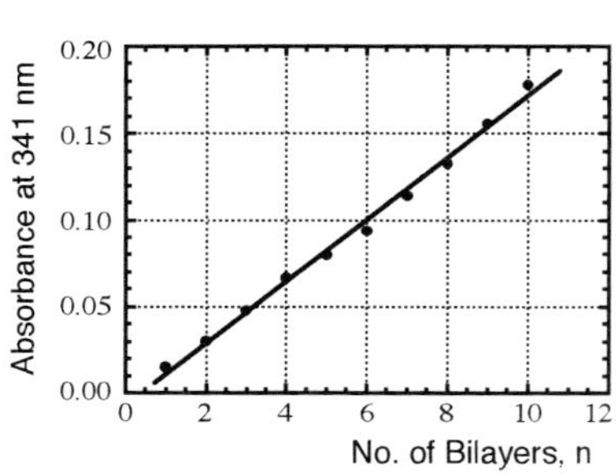

Fig. 2: UV absorbance at 341 nm vs. number of bilayers, BA-12/PVS.

1.2. Morphology of Multilayer Assemblies

The structure of the multilayer assemblies deposited on silicon wafer was probed by ellipsometry. We obtained the thickness of 136±4 Å for the precoated 5 bilayers of PAH/PVS and that of 358±12 Å for 10 bilayers of BA-12/PVS. Since the true refractive indices are unknown, we assumed a film refractive index to be 1.54 to analyze the ellipsometry data. The average thickness of each polymeric layer is 13.6±0.4 Å that is in the same order of magnitude as once reported for a polyelectrolyte monolayer adsorbed on silicon substrate[26)]. According to the measurement, we can calculate the thickness of a BA-12 layer to be 22.2±1.6 Å, assuming the thickness of a PVS layer to be 13.6±0.4 Å which was estimated from the determined total thickness of the 5 bilayers of PAH/PVS.

We have also attempted to estimate the orientation of azobenzene chromophore in the BA-12/PVS film based on the polarized UV spectra obtained at 45° incidence angle. The estimation was made by using the optoelectric equations reported by Vandevyver et al.[27)]. The infinite dilution hypothesis was applied, and the $\pi\pi^*$ transition dipole of trans-azobenzene is assumed to be aligned along a long molecular axis. The refractive indices of fused silica and BA-12 are taken to be 1.47 and 1.43, respectively[27)]. The average tilt angle of the azobenzene was computed to be 63.7° with respect to the surface normal.

From a molecular modeling calculation (Hyperchem), the molecular length of BA-12, in fully-stretched form, is estimated to be 46.2 Å. Based on this, the average tilt angle of a BA-12 molecule with respect to the surface normal is 61.3° from the geometrical calculation. This is approximately consistent with the value obtained from the measurement. The fact indicates that the BA-12 seems therefore to adopt a slightly parallel stance, when the bolaamphiphiles are adsorbed on the oppositely-charged polymer surface.

1.3. Photoinduced Cis-Trans Isomerization of Azobenzene

Fig. 3(a) shows the absorption spectra of the multilayer assembly composed of BA-12/PVS bilayers, before and after various UV irradiation times. The spectra are recorded with the linearly polarized probe light oriented parallel to the irradiating UV light polarization. The maximum absorbance of molecularly dispersed azobenzene at 358 nm is blue-shifted to 341 nm in the dilute aqueous solution and the SA film, indicating the interaction of the chromophores in BA-12 forming H-aggregates. The trans-azobenzene chromophores should be arranged "face-to-face" in dimers or higher aggregates. The process can be described qualitatively by the exciton model introduced by Kasha[28)]. The change in spectral shape and the existence of the isosbestic points at 308 nm and at 440 nm clearly demonstrate the trans-cis photoisomerization reaction. The kinetics of the photoisomerization is characterized by a uniform decrease of absorbance at the π-π^* band of trans isomer (341 nm) and an increase of the n-π^* band of cis isomer (450 nm). The absorption band at 320 nm observed upon UV irradiation can be assigned to the π-π* transition of the cis isomer.

When the photostationary state is attained after 20 min. of UV irradiation, the cis-trans reverse isomerization that is induced by irradiation with unpolarized blue light (450 nm), can bring the azobenzene back to the initial state. In Fig. 3(b), the recovery of the trans state and the existence of the isosbestic points clearly demonstrate that the blue light induced cis-trans back isomerization. The trans state prior to UV irradiation was restored after only $2^1/_3$ min. of blue light irradiation. However, it should be also noted that after first trans-cis-trans cycle, the absorbance maximum at 341 nm was drastically shifted to that of the molecularly isolated chromophore at 358 nm. This reveals that the H-aggregate of the azobenzene in the multilayer assemblies was destroyed to the molecularly dispersed state by the isomerization cycle.

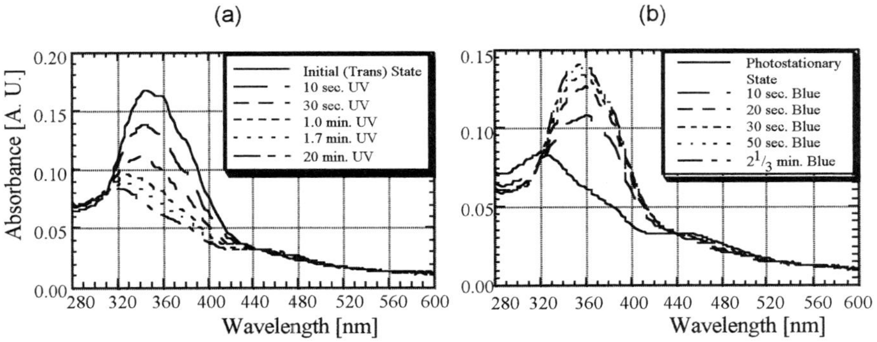

Fig. 3: UV/Vis-absorption spectra of a multilayer assembly composed of BA-12 and PVS on a fused silica precoated with 5 bilayers of PAH and PVS: (a) before and after various timed doses of linearly polarized UV (360 nm) irradiation, (b) before and after various amounts of unpolarized blue (450 nm) irradiation. The polarized light was linearly polarized and oriented parallel to the exciting UV irradiation.

In order to study reversible photochemical switching between the two isomers, the multilayer assemblies containing 10 bilayers of BA-12 and PVS were irradiated with UV (360 nm) and blue light (450 nm) for trans-cis and cis-trans isomerization, respectively. In the first trans-cis-trans isomerization cycle, the photostationary state reached by irradiating the sample with the UV light for 20 min. However, after first one cycle, the irradiation time for the photostationary state in the following isomerization cycles decreased markedly from 20 min. to 30 sec. The difference in switching times required for trans-cis isomerization can be ascribed to the increased free volume in the multilayer that is created by the first trans-cis-trans cycle. Since the cross-sectional area of the cis isomer is larger than that of the trans isomer, free volume is considered to be necessary for the molecule to photoisomerize. Hence, the photochemical behavior directly allows conclusions to be drawn about the free volume available in the multilayer assemblies. The trans-cis and cis-trans photoisomerization were repeated in 10 cycles without appearance of any fatigue in the multilayer film, as shown in Fig. 4.

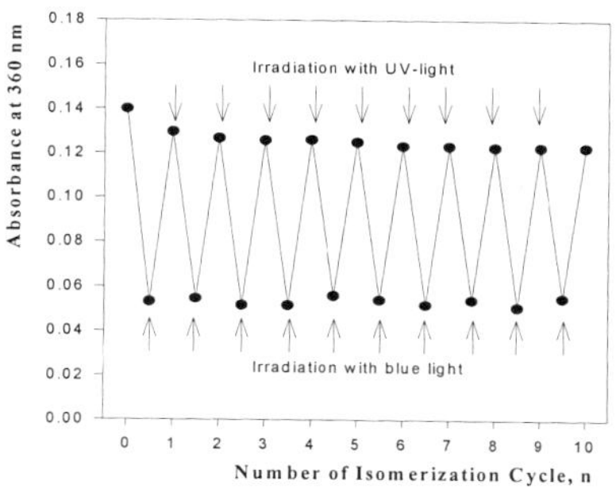

Fig. 4: The change of the maximum absorbance at 360 nm after 30 sec of UV and blue light irradiation for trans-cis and cis-trans isomerization, respectively.

2. Alternating Multilayer Assemblies of Polyelectrolytes and Ionene-type Polyion

2.1. Preparation and Structure of Multilayer Assemblies

The electrostatic SA technique is now shown to allow the multilayer assemblies with excellent control over thickness and molecular architecture. However, especially in the case of the polyelectrolyte multilayer, we rarely encounter the assembly that consists of internal orderings in each polymer monolayer. Laschewskey et. al. reported about layered nanostructures with an internal ordering of the assembly using liquid crystalline ionomers[29]. In this article, we will show that an internal ordering of the multilayer assemblies can be achieved by using an ionene oligomer PAZ-12 bearing azobenzene unit on main-chain. The molecular weight of the ionene was about 13 estimated by using the characteristic signals of its nmr spectrum.

Kunitake et. al.[25] reported that stable molecular membranes are formed in an aqueous solution spontaneously from ionene-type amphiphiles (see Scheme II). The ionene amphiphile is an oligomer composed of a repeating unit with a bolaamphiphile. Taking the work into account, we designed an ionene polymer containing an azobenzene chromophore connected with C_{12} alkyl chains, assuming that stable molecular membranes form from ionene oligomers in a dilute solution, as shown in the Scheme II.

Scheme II

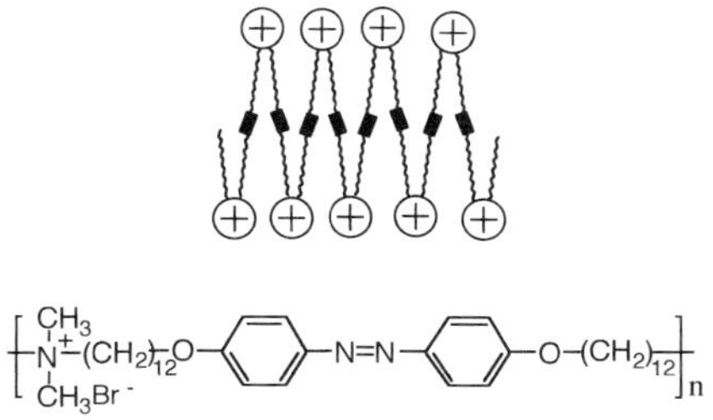

PAZ-12, **2**

We have constructed a SA multilayer film composed of a cationic ionene PAZ-12 and an anionic polyelectrolyte PVS on a fused silica precoated with five bilayers of PAH and PVS, as shown schematically in Fig. 5. Photoactive PAZ-12 layer exists between two polyelectrolyte layers in order to allow for a photoisomerization to take place.

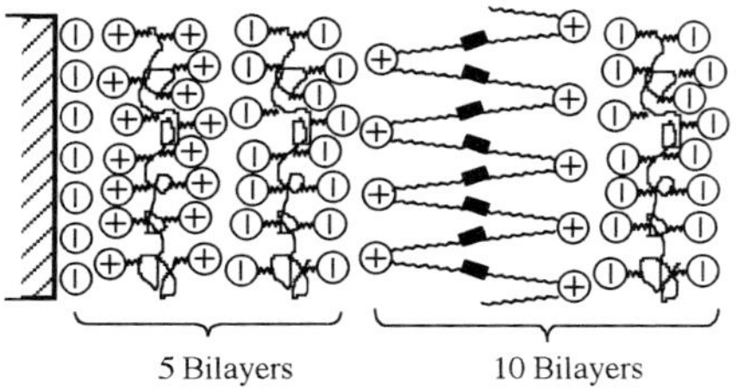

Fig. 5: Idealized schematic structure of the multilayer assemblies composed of the cationic ionene PAZ-12 and the anionic polyelectrolytes PVS.

By UV/Vis-spectroscopy we have also demonstrated that the consecutive adsorption of layers is a stepwise and regular process. The regular formation of each PAZ-12 layer is illustrated by a linear correspondence between the number of bilayers deposited and the absorbance of the multilayer film up to at least 10 bilayers, as shown in the insets of Fig. 6. The points can be fitted with a linear fit, yielding an average optical density of 0.011 per layer.

The orientation of azobenzene moiety in the PAZ-12/PVS film was estimated by referring to the optoelectric equations reported by Vandevyver et. al. [27] based on the polarized UV spectra obtained at 45° incidence angle. The infinite dilution hypothesis was applied, and the $\pi-\pi^*$ transition dipole was assumed to be directed along molecular axis of azobenzene. The refractive indices of fused silica and PAZ-12 layer were taken to be 1.47 and 1.43[27]. On this basis, the average tilt angle of the azobenzene moiety with respect to the substrate normal was calculated to be 49.1°.

The thickness of a PAZ-12 layer sandwiched between the polyelectrolyte layers was estimated by using an optical ellipsometry. According to the measurement, we have obtained 24.8±2.1 Å for the thickness of a PAZ-12 layer based on 13.6 Å for that of a PVS monolayer, assuming a film refractive index to be 1.54 in

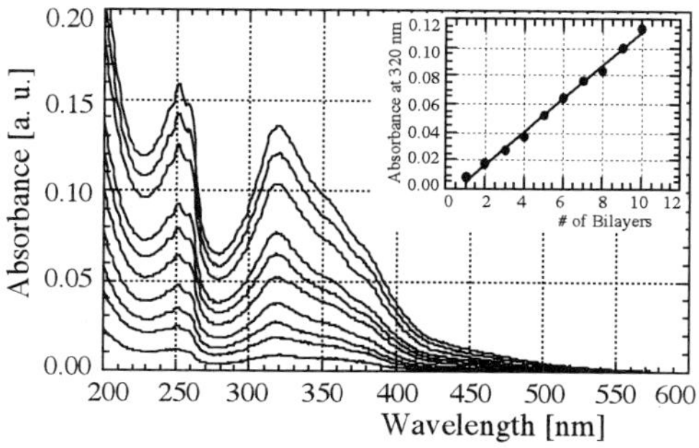

Fig. 6: Optical absorption spectra of self-assembled multilayer built up by alternate adsorption of PAZ-12 and PVS. Samples differ in the number of dipping cycles applied. In the Insets, the maximum absorbance of PAZ-12 in the respective multilayer is plotted versus number of bilayers of PAZ-12 and PVS.

analyzing the ellipsometric data. Anticipated from the ellipsometry data, the azobenzene moiety should adopt a rather parallel orientation. The thickness of a PAZ-12 layer (24.8 Å) deviates slightly from the theoretical value (28.2 Å) that is obtained from the simple geometrical calculation with the molecular length of the repeat unit (46.2 Å) and the measured tilt angle of the chromophore (49.1°). However, more work has to be done to obtain a more complete picture of PAZ-12/PVS multilayers.

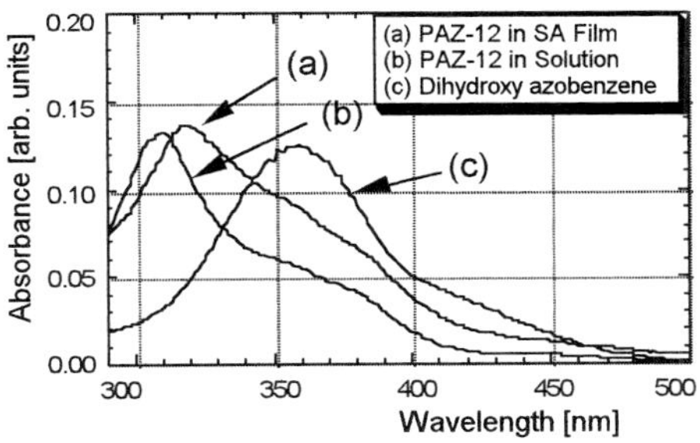

Fig. 7: Comparison of the UV/Visible absorption spectra of PAZ-12 in solution (a), and SA film (b), and the normalized spectra for 4,4'-dihydroxy azobenzene in solution (c).

2.2. Photoisomerzation of Multilayer Assemblies

Fig. 7 compares UV/Visible spectra of the SA 10 bilayers of PAZ-12 and PVS on a fused silica (a), PAZ-12 in DMSO/H_2O (10/3) (b), and molecularly dispersed 4,4'-dihydroxy azobenzene (c). The maximum absorbance of the molecularly dispersed chromophore at 358 nm is drastically blue-shifted to 310 nm in the solution of PAZ-12. The symmetric shape of the π-π* absorption band becomes asymmetic and narrow. Both facts indicate a strong excitonic interaction of the aromatic chromophores forming H-aggregates. The UV absorbance depending temperature indicates that the formed H-aggregate of PAZ-12 in solution are destroyed to the molecularly dispersed state around 90°C and reformed after cooling to r.t. The multilayer assembly had a strong band with maximum at 320 nm, indicating highly crystalline layers of H-aggregated chromophores in parallel alignment. We found that the absorbance of the PAZ-12 in SA film differs essentially not from that in solution. The fact supports our assumption that the polymeric multilayer assembly with the internal orderings can be realized by transferring the monolayer membrane formed in solution on substrate. Upon irradiation the azobenzene undergoes trans-cis photoisomerization and a wavelength-dependent steady state is established between rod-like trans isomers and bent cis isomers. Fig. 9 shows the change of the absorption spectrum of the SA film bearing 10 bilayers of PAZ-12 and PVS on irradiation with blue light (450 nm). The photoisomerization occurs relatively large within 12 min. (ca. 95% with regard to the photostationary state) and the stationary state was established by irradiating the film for ca. 48 min. The trans-cis isomerization was revealed by the decrease and the increase of the absorption at 320 nm and at 447 nm that can be ascribed to the π-π* band

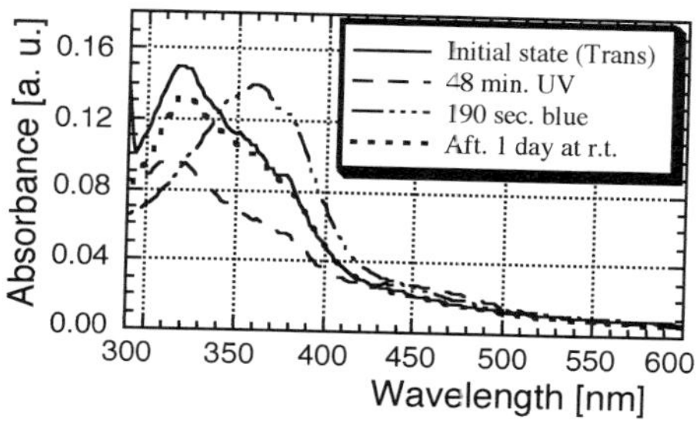

Fig. 8: UV/Visible absorption spectra of the multilayer assemblies composed of 10 bilayers of PAZ-12 and PVS; (a) initial (trans) state, (b) cis state after UV irradiation, (c) trans state after blue irradiation, (d) trans state after storage at r.t. for one day.

of the trans isomers and the n-π* band of the cis isomers, respectively. The subsequent irradiating with unpolarized blue light induces cis-trans back isomerization with the rapid increase of the maximum absorbance at 357 nm. Once again, a clear isosbestic point at 422 nm was observed during the reverse cis-trans isomerization. The photostationary state was attained by irradiating the sample with the blue light for ca. 190 sec. An important observation in this trans-cis-trans cycle, however, was that the maximum absorbance in SA film at 320 nm is shifted to 357 nm with the formation of the symmetric shape of π-π* band. The red-shifted absorption spectrum returned slowly to the original state and to the almost same state after one day. From the spectral data presented above, it is clear that the aggregates are destroyed to the molecularly dispersed state with the disordering of main-chain by bent cis isomers resulted from trans-cis photoisomerization process, and slowly reformed again with realignment of the chains in the solid film by keeping it at r.t. for one day.

In order to study reversible photochemical switching between the two isomers, the multilayer assemblies containing 10 bilayers of PAZ-12 and PVS were irradiated with UV (360 nm) and blue light (450 nm) for trans-cis and cis-trans isomerization, respectively. In the first trans-cis-trans isomerization cycle, the photo-stationary state reached by irradiating the sample with the UV light for ca 48 min. However, after first one cycle, the irradiation time for the photostationary state in the following isomerization cycles decreased markedly from 48 min. to 30 sec. The difference in switching times required for trans-cis isomerization can be ascribed to the increased free volume in the multilayer that is created by the first trans-cis-trans cycle, as observed in the sample of

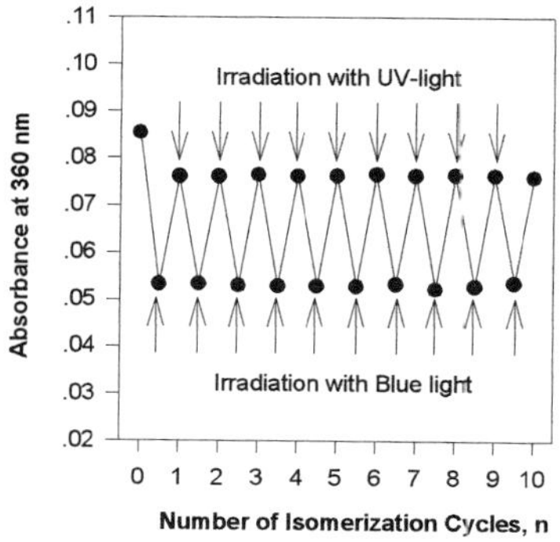

Fig. 9: The change of the maximum absorbance at 360 nm after 30 sec of UV and blue light irradiation for trans-cis and cis-trans isomerization, respectively.

BA-12 and PVS. The trans-cis and cis-trans photoisomerization were repeated in 10 cycles without appearance of any fatigue in the multilayer film, as shown in Fig. 9.

Conclusions

We have shown that multilayer assemblies of a new ionene oligomer bearing azobenzene unit on main chain and an anionic polyelectrolyte can be prepared by electrostatic layer-by-layer adsorption method. The influence of the supramolecular order in the multilayer assemblies on the photoisomerization was discussed with varying the component of the photoactive layers sandwiched between oppositely-charged polyelectrolytes layers from a small molecule, bolaamphiphile BA-12 to an ionene oligomer PAZ-12. We have observed that the chromophore interaction in PAZ-12 is stronger than that in BA-12 in solution and SA film (λ_{max} = 310 nm or 320 nm compared to 341 nm), although the aggregate formation should be considered to be hindered by the strains of the polymer chains.

By the photoisomerization, it has been found that after first one cycle, the irradiation time for the photostationary state in the following isomerization cycles decreased markedly from 20 min. and 48 min. to 30 sec. for BA-12/PVS and PAZ-12/PVS multilayers, respectively. Also, we could demonstrate that the trans-cis and cis-trans photoisomerization were repeated in 10 cycles without appearance of any fatigue in the multilayer film. The differences in switching times required for trans-cis isomerization can be ascribed to the increased free volume in the multilayer that is created by the first trans-cis-trans cycle. Finally, we believe that the molecular assemblies are useful as a medium for reversible photochemical switching and optical storage of information

Acknowledgment. This paper was supported by NON DIRECTED RESEARCH FUND, Korea Research Foundation and the Ministry of Science and Technology, Korea. Authors express appreciation to Prof. Kwan Kim in Seoul National University for the stimulated discussions and allowing the use of ellipsometer.

References

[1] G. Decher and J. D. Hong, Makromol. *Chem. Macromol. Symp.* **46**, 321 (1991)
[2] G. Decher, J. D. Hong, and J. Schmitt, *Thin Solid Films* **210/211**, 831 (1992)
[3] Y. Lvov, G. Decher, and G. Sukhorukov, *Macromolecules* **26**, 5396 (1993)
[4] E. Donth, D. Walther, V. N. Shilov, E. Knippel, A. Budde, K. Lowack, C. A. Helm and H. Möhwald, *Langmuir* **13**, 5294 (1997)
[5] Y. Lvov, K. Ariga, I. Ichinose, and T. Kunitake, *J. Am. Chem. Soc.* **117**, 6117 (1995)

[6] E. R. Kleinfeld, G. S. *Science* **265**, 370 (1994)
[7] S. W. Keller, H. N. Kim, T. E. Mollouk, *J. Am. Chem. Soc.* **116**, 8817 (1994)
[8] Y. Lvov, K. Ariga, I. Ichinose, and T. Kunitake, *Langmuir* **12**, 3038 (1996)
[9] N. A. Kotov, I. Dékány, J. H. Fendler, *J. Phys. Chem.* **99**, 13065 (1995)
[10] D. L. Feldheim, K. C. Crabar, M. J. Natan, T. E. Mallouk, *J. Am. Chem. Soc.* **118**, 7640 (1996)
[11] J. Schmitt, G. Decher, W. J. Dressik, S. L. Branduo, R. E. Geer, R. Shashidhal, J. M. Calvert, *Adv. Mater.* **9**, 61 (1997)
[12] R. G. Freeman, K. C. Grabar, K. J. Allison, R. M. Bright, J. A. Davis, A. P. Guthrie, M. B. Hommer, M. A. Jackson, P. C. Smith, D. G. Walter, M J. Natan, *Science* **267**, 1629 (1995)
[13] T. Cooper, A. Campbell, R. Crane, *Langmuir* **11**, 2713 (1995)
[14] D. Yoo, J. –K. Lee, M. F. Rubner, *Mater. Res. Soc. Symp. Proc.* **413**, 395 (1996)
[15] X. Zhang, M. Gao, X. Kong, Y. Sun, J. Shen, *J. Chem. Soc., Chem. Commun*, 1005 (1994)
[16] K. Araki, M. J. Wagner, M. S. Wrighton, *Langmuir* **12**, 5393 (1996)
[17] B. Sellergren, A. Swietlov, T. Amebrant, K. Unger, *Anal. Chem.* **68**, 402 (1996). G. Mao, Y. Tsao, M. Tirrell, H. T. Davis, *Langmuir* **9**, 3461 (1993). X. Zhang, M. Gao, X. Kong, Y. Sun, J. Shen, *J. Chem Soc., Chem. Commun.* 1055 (1994). F. Saremi, B. Tieke, *Adv. Mater.* **7**, 378 (1995). F. Saremi, E. Maassen, B. Tieke, *Langmuir* **11**, 1068 (1995). F. Saremi, G. Lange, and B. Tieke, *Adv. Mater.* **8**, 923 (1996)
[18] K. Ariga, Y. Lvov, and T. Kunitake, *J. Am. Chem. Soc.* **119**, 2224 (1997)
[19] J. H. Cheung, A. F. Fou and M. F. Rubner, *Thin Solid Films* **244**, 958 (1994). C. Fou, O. Onitsuka, M. Ferreira, and M. F. Rubner, *J. Appl. Phys.* **79**, 7501 (1996). J. -D. Hong, D. S. Kim, K. H. Char and J. I. Jin, *Synth. Metals* **84**, 815 (1997)
[20] G. Mao, Y. Tsao, M. Tirrell, H. T. Davis, V. Hessel, H. Ringsdorf, *Langumir* **11**, 942 (1995)
[21] J. -D. Hong, E. S. Park, and A. L. Park, *LB8 Conference*, Asilomar, 1997
[22] F. Saremi and B. Tieke, *Adv. Mater.* **10**, 388 (1998)
[23] J. -D. Hong, E. S. Park, and A. L. Park, *Bull. Korean Chem. Soc.* **19**, 1156 (1998)
[24] A. Laschewsky, B. Meyer, E. Wischerhoff, X. Arys, P. Bertrand, A. Delcorte, A. Jonas, *Thin Solid Films* **284-285**, 334 (1996)
[25] T. Kunitake, A. Tsuge, K. Takarabe, *Polymer Journal* **17**, 633 (1985)
[26] Y. Lvov, G. Decher, H. Möhwald, *Langmuir* **9**, 481 (1993)
[27] M. Vandevyver, A. Barraud, P. Maillard, Raudel-Teixier, C. Gianotti, *J. Colloid Interface Sci.* **85**, 571 (1982)
[28] M. Kasha, *Radiat. Res.* **20**, 55 (1963)
[29] D. Cochin, M. Passmann, G. Wilbert, R. Zentel, E. Wischerhoff, A. Laschewsky, *Macromolecules* **30**, 4775 (1997)

Macromol. Symp. ***142,*** *133–143 (1999)*

Perfectly Branched and Hyperbranched Poly(ether amide)s

Thomas Huber, Brigitte Voit*, Dieter Wolf

Institut für Polymerforschung Dresden e.V., Hohe Strasse 6, D-01069 Dresden, Germany

SUMMARY: Poly(ether amide) dendrimer segments based on 5-(2-aminoethoxy)-isophthalic acid have been synthesized using the convergent approach and mild reaction conditions. The monodendrons were combined to dendrimers via different core molecules and the resulting properties of the molecules were compared to those of the monodendrons. The melt polycondensation of the AB_2 monomers with analog structure resulted first in undefined polymer structures. However, by thermal polyaddition reaction of suitable functionalized oxazoline monomers, well defined hyperbranched poly(ether amide)s with phenol terminal groups could be synthesized.

Introduction

Dendrimers have become a very popular topic in polymer chemistry in the last years due to the beauty of their perfectly branched structure in combination with unusual properties, such as high solubility, low viscosity, high functionality, and different reactivity compared to linear polymers. Using the divergent (from the inside out) or the convergent (from the outside in) approach a very large number of different dendrimers[1)] has been synthesized, including poly(amido amine)s[1)], poly(ether amides)[2,3)], poly(ester amide)s[2)], poly(amide)s[2-5)], and poly(aramide)s[5-8)]. The amide bond is of high interest in the dendrimer synthesis due to its chemical and thermal stability which reduces the danger of side reactions, but also due to the large amount of knowledge accumulated on the stepwise synthesis in peptide chemistry and regarding protective group strategies. Even solid-phase synthesis of dendritic polyamides has been attempted in analogy to the Merrifield approach[9)]. In addition, the resemblance of polyamide dendrimers with biopolymers like peptides results in an increasing effort to apply these dendrimers in biochemistry or medicine[1,10,11)].

Hyperbranched polyamides, synthesized in a one step reaction from AB_2 monomers and with a less defined structure than dendrimers but with similar properties, are very interesting from

 CCC 1022-1360/99/$ 17.50+.50/0

the materials point of view e.g. for blends or coatings. Linear polyamides are commonly available as materials with high modulus due to semicrystallinity or high glass transition temperatures (T_g). However, the semicrystallinity and the strong tendency to form hydrogen bonding causes also low solubility and high melt viscosity which limits the processing. A highly branched structure as in dendrimers or hyperbranched polymers, which usually leads to amorphous materials with excellent solubility, might improve the processing of polyamides and therefore, can result in new applications for polyamides. First results on highly branched polyaramides[12-15] give evidence that high molar masses can be achieved in combination with good solubility and T_g above room temperature. Our goal is the synthesis of perfectly branched dendrimers and analog hyperbranched polyamides with glass transition temperatures above room temperature. An additional requirement is the presence of reactive functional groups on the dendrimer surface which should allow further modification reactions.

Results and Discussion

Poly(ether amide) dendrimers

Poly(ether amide) dendrimer segments up to generation 4 have been synthesized based on 5-(2-aminoethoxy)-isophthalic acid **1** as building block and using the divergent approach towards dendrimers[16,17]. A similar approach towards aliphatic-aromatic poly(amide ether) dendrimers but with amino terminal units based on 3,5-(2-aminoethoxy)-benzoic acid has been reported recently by Liskamp et al.[18]. For the repetitive synthesis of the monodendrons of different generation the starting material **1** (obtained via Mitsunobu[19] reaction from 5-hydroxyisophtalic acid) had to be converted into the amino protected compound **2** and the acid protected analogon **3**. One equivalent of **2** was combined with two equivalents of **3** to a monodendron of generation 2 (**G2-BOC**) under mild reaction condition using different coupling reagents like DCC, EDC, HATU, as applied in peptide synthesis. Also very effective was the use of carbonyldimidazole (CDI) which led to the *in situ* formation of the activated intermediate **4** (scheme 1). The BOC-protection group of **G2-BOC** was removed and two equivalents of **G2-Amine** were reacted with one equivalent of **2** to yield **G3-BOC** and so on up to generation 4.

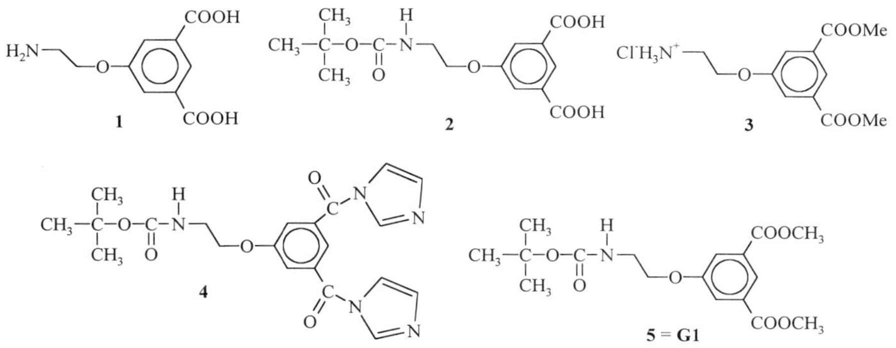

Scheme 1

The structure of the molecules as well as the purity and the theoretical molar masses could be verified by NMR analysis, MS-MALDI-TOF and elemental analysis. Especially MS-MALDI-TOF proved to be very useful for the determination of side reactions or incomplete conversion whereas small deviations in the perfect structure were no longer detectable in the NMR e.g. for **G4**. Even though the perfect structure of the monodendrons could be proven, the removal of small amounts of the activating agents or of the applied solvents was nearly impossible for the larger structures which is an indication for the increasing globular and compact structure of the molecules. Computer modeling and molecular dynamics experiments proved the globular structure[16], but also that no sterical hindrance occurs and further growth should be possible. First attempts to synthesize monodendron **G5** were successful with regard that a compound of the expected molar mass could be isolated but only

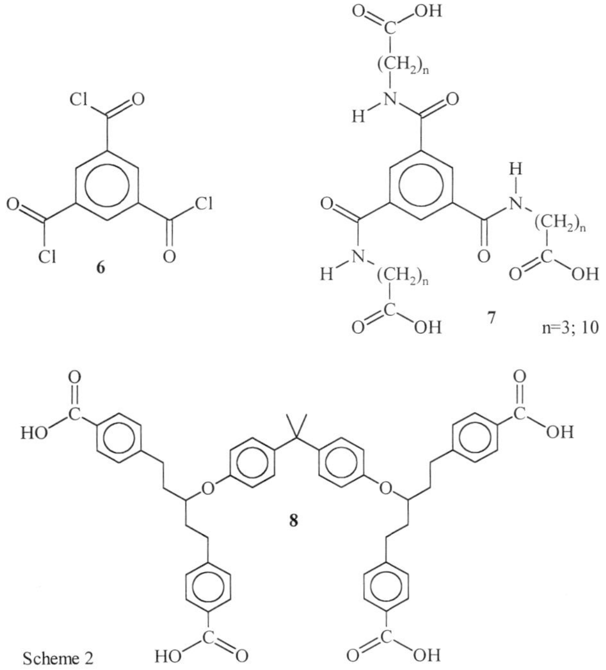

Scheme 2

in low yield.

As multifunctional cores the commercially triacid chloride **6** as well as the more flexible structures **7** (synthesized from **6** via amidation with ω-aminoundecanoic acid methyl ester and subsequent hydrolysis of the ester using potassium carbonate or sodium hydroxid) were applied (Scheme 2). In addition, the tetrafunctional compound **8** was synthesized by Mitsunobu etherification of 1,5-bis(4-methoxycarbonylphenyl)-3-pentanol[20] with bisphenol A and hydrolysis of the methyl ester functions. The very rigid core **6** was reacted with monodendrons **G1** to **G3** but already with generation 2 no complete reaction could be reached due to sterical hindrance at the core. The core molecule **8** – converted to the tetraacid chloride - was already successfully reacted with **G2** monodendrons leading to tetradendrons with a rather large, aliphatic-aromatic core area. The structure could be fully verified by ^{13}C and ^{1}H NMR analysis as well as SEC but further characterization still has to be performed.

The core molecule **7**, with the very flexible spacers, does not lead to sterical hindrance and complete attachment of **G1** to **G3** monodendrons was possible. The structure of the tridendron with **G3** (**9**) is demonstrated in Scheme 3.

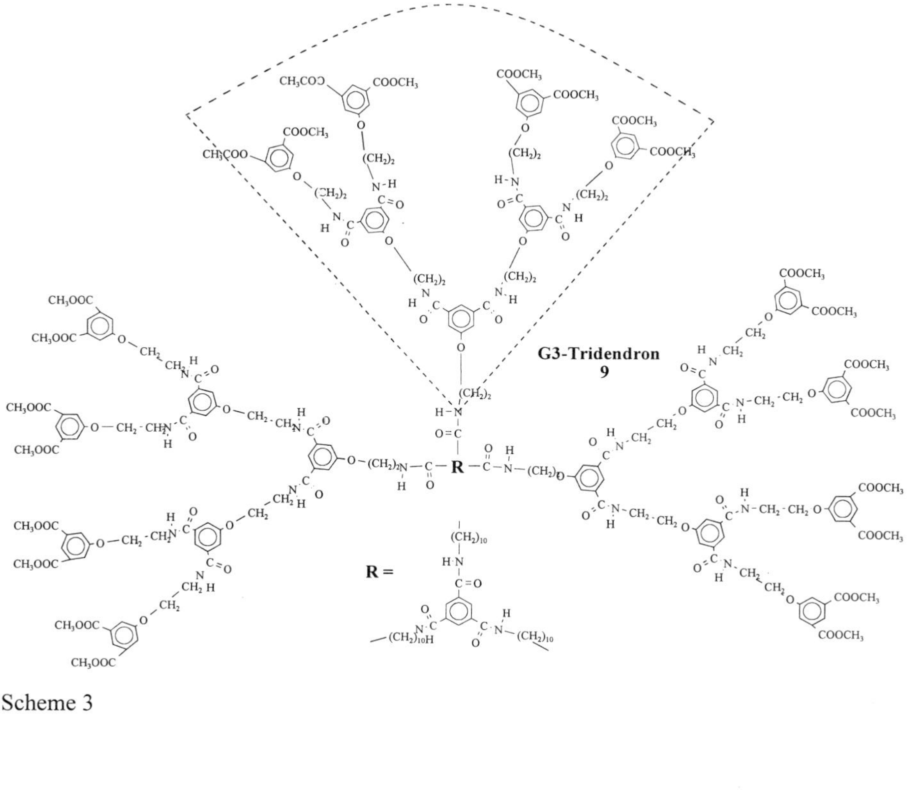

Scheme 3

The monodendrons of generation 1 to 3 exhibit in bulk still some crystallinity but the enthalpy value of the melting endotherm in the DSC measurements decreases with increasing generation number. Already **G2-BOC** shows a well distinguishable glass transition at 44 °C which shifts to to about 95 °C for **G3** and **G4**. No crystallinity was observed for **G4** and all other tri- and tetradendrons. Despite of the flexible core molecule the glass transition temperature of **9** was similar to that of **G4-BOC**. However, the solubility behavior changes strongly from **G4** to the tridendron: the monodendron is only soluble in DMF or DMAc and in very low concentration in THF, but **9** exhibits excellent solubility in THF and even in chloroform and methylene chloride.

Table 1: Results for the monodendrons **G1** to **G4** and their products with core **9**

compound	$M_{theor.}$ g/mol[1]	$M_{n(SEC)}$ g/mol[2]	$\eta_{inh.}$ dL/g[3] (c = 0.2g/L)	T_m °C[4]	T_g °C[5]
G1-BOC (**5**)	353	460	-[6)]	82	-
G2-BOC	796	790	-[6)]	113	44
G3-BOC	1680	1370	0.061	111	97
G4-BOC	3450	2350	-[6)]	-	93
G3-tridendron (**9**)	5446	20500	0.250	-	109

[1)] theoretical molar mass M_n could be verified for all compounds by MS-MALDI-Tof

[2)] by SEC in THF or DMAc (**9**) calibration with linear polystyrene; [3)] measured in DMF; by DSC, 20K/min, 1st heating run; [4)] by DSC, 20K/min, 2nd and 3rd run; [6)] not measured

As described above, the molar masses were determined successfully by MS-MALDI-TOF. SEC analysis of the monodendrons in THF applying linear polystyrene calibration resulted for the larger molecules in a increasing deviation of the measured molar mass values from the theoretical values which is in agreement with the increasing globular shape of the molecules (Table 1). The SEC molar mass of **9** has been performed in DMAc as a typical solvent for polyamides, again using polystyrene calibration. Surprisingly, the SEC molar mass of **9** is much larger than the theoretical molar mass which is mainly due to the limitation of the polystyrene calibration in polar solvents. However, aggregation of the molecules can also not be excluded. Solution viscosity measurements show a remarkable increase in the inherent viscosity from **G4** to **9** with adding salt to the solvent having no effect. MS-MALDI-TOF, however, clearly exhibit the expected molecule peak for **9**, even though a small impurity of unreacted **G3** and of a side reaction product of the core molecule can also be detected. Further purification of this dendrimer is still in progress.

The methylester end groups can be split off without degradation of the internal amide bonds by selective hydrolysis yielding carboxylic acids, when mild reaction conditions are applied and potassium carbonate is used as base. Thus, reactive functions can be created on the dendrimer surface which are available for further modification reactions.

Hyperbranched poly(ether amide)s

Jikei et. al.[15] described the synthesis of fully aromatic poly(ether amide)s and a first approach towards hyperbranched aliphatic-aromatic poly(ether amide)s of an analogous structure of the above described dendrimers has be published previously[17] by our own group. Both approaches use conventional polycondensation reactions of acids and amines to form the hyperbranched structure. However, using the building block **3** or **5** in a high temperature melt polycondensation process did not result in clean hyperbranched polyamides up to now. Side reaction like partial hydrolysis of the terminal ester groups and alkylation of the amide functions occurred.

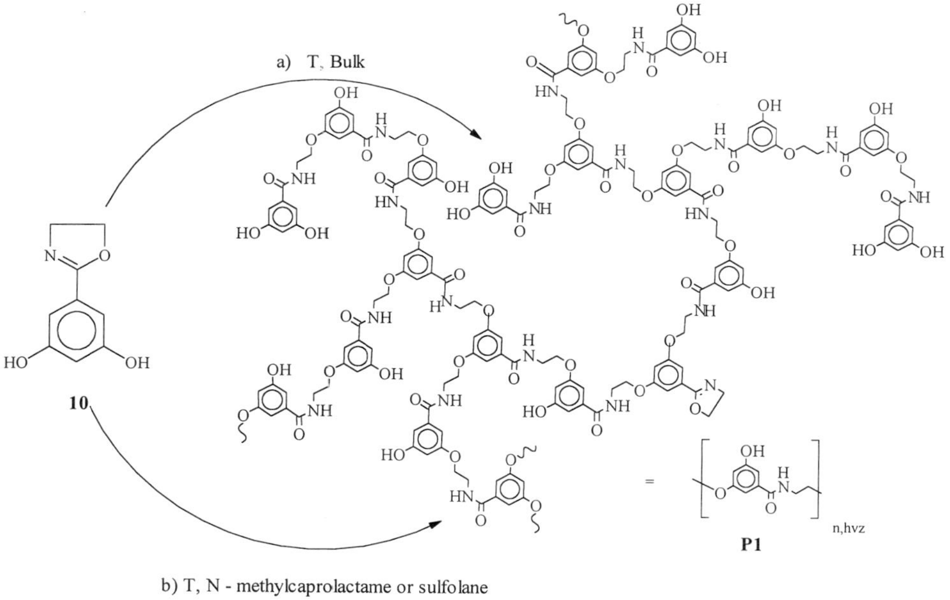

Scheme 4

Therefore, a different approach towards hyperbranched polyamides has been undertaken. The nucleophilic ring opening reaction of 2-oxazolines with carboxylic acids, thiols and phenols leads to esteramides[21-23], thioetheramides[24,25] and etheramides[26,27]. The synthesis of linear poly(ether amide)s in melt and solution of monomers containing a 2-oxazoline and a phenolic group was described by Mülhaupt et al.[28]. We found that the polymerization of an AB_2 monomer with two phenolic groups and one oxazoline function leads to well defined hyperbranched poly(ether amides)[29] (Scheme 4). The structure of the repeating units corresponds well with those of the dendrimers described above with the exception that phenol end groups are obtained instead of methylester or acid functions.

The reactions proceeds best in N-methylcaprolactame as solvent at 190 °C but reaction in bulk at 220 °C was also possible. The products are highly soluble in polar solvents like DMF, DMAc, and DMSO. The ^{1}H and ^{13}C NMR analysis by means of two dimensional experiments allowed the exact assignment of all signals of the different structural units in the polymer. The ratio of linear, dendritic and terminal units could be determined and a degree of branching of 50% was calculated which is in agreement with a fully statistical branching process. In contrast to observation on linear poly(ether amides) synthesized from p-hydroxyphenyl-2-oxazoline[26], no side reactions could be detected in the hyperbranched system when water was eliminated completely from the reaction medium and the starting materials.

By optimization of the reaction conditions, medium and time, hyperbranched poly(ether amide)s up to a molar mass M_n of 47,500 g/mol could be synthesized. This polymer exhibits an $\eta_{inh.}$ of only 0.305 dL/g (c = 0.2 g/dL) in DMF and a product with M_n = 13500 g/mol of 0.079 dL/g, whereas a linear analog oligomer synthesized from m-hydroxyphenyloxazoline of a molar mass of only 1700 g/mol is much less soluble in DMF and has a η_{inh} of already 0.098 dL/g (c = 0.2 g/dL).

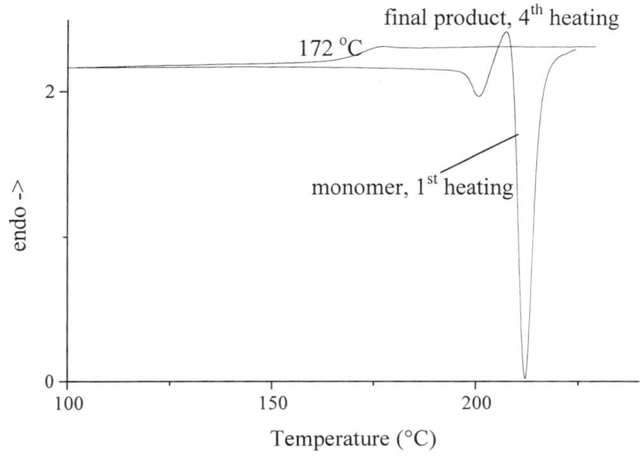

Fig. 1: DSC trace of the AB_2 monomer **10** and the resulting hyperbranched polymer (heating rate: 10K/min)

The thermal reaction of the monomer **10** can be followed by DSC analysis (Fig.1). In the first heating cycle the thermal addition reaction can be observed as a strong exothermic peak which starts directly after melting of the compound is induced. The polyreaction temperature of 190 °C in solution and 220 °C in bulk fits directly into this melting/reaction area. After the first heating cycle the product can be further thermal treated in the DSC experiment by several heating cycles until no more changes occur in the curve (4th heating cycle). The T_g of the resulting polymer can be observed already in the second heating cycle and it shifts up to 172 °C in final heating. Glass transitions temperatures between 170 and 176 °C are also observed for the polyreaction products carried out in the reaction flask.

Thus it was possible to synthesize perfectly branched poly(ether amide) dendrimers and hyperbranched polymers with similar structure. The products are more stable towards hydrolysis than e.g. hyperbranched polyesters, have glass transition temperatures far above room temperature, show high solubility and low solution viscosities.

Experimental Part

Details to the synthesis of the monodendrons G1 to G4 can be found in ref.[16)]; the polymerization of **10** is described in ref.[29)]. 1,5-Bis(4-methoxycarbonylphenyl)-3-pentanol was synthesized following a patent literature[20)]. All products have been characterized by SEC (dendrimers: in THF or DMAc, polystyrene calibration; hyperbranched polymers: in DMAc/LiCl/H_2O, polystyrene calibration; UV- and RI detection). The molar mass of the products has been further verified by MS-MALDI-TOF (HP G2025A MALDI-TOF, measurements in a THF / 2,5-dihydroxybenzoic acid matrix with K^+ - modifier using a laser energy of 5.79 μJ). 500.13 MHz ^{1}H-NMR spectra and 125.74 MHz ^{13}C-NMR spectra were recorded with a Bruker DRX 500 NMR.

Synthesis of the core molecules

Trimesinic acid-tris-(11-carboxyundecylamide) **7**

ω-Aminoundecanoic acid was converted to the methyl ester via standard esterification procedure. The ester (6.4g, 30 mmol) and trimesylchloride (2g, 7.5 mmol) were dissolved in 100 mL dichloromethane. The solution was cooled to 0°C and triethylamine (10 mL, 72 mmol) was added. After 24 h stirring at room temperature the reaction was stopped by addition of 50 mL water and 1g $NaHCO_3$ and a small amount of tributyl ammonium bromide. The organic phase was extracted with diluted HCl, dried over Na_2SO_4, and the product was isolated by removal of the solvent. The methyl ester function were hydrolyzed using aqueous NaOH or potassium carbonate. The triacid was isolated by adjusting the pH to 2-3 and subsequent filtration. Yield:3.1 g (55% th)

^{1}H NMR (d$_6$-DMSO, δ in ppm): 1.30, 1.56, 2.23, 3.32, 8.47, 8.69,

^{13}C NMR (d$_6$-DMSO, δ in ppm): 25.3, 7.3, 29.4, 29.6, 29.7, 29.8, 29.9, 34.5, 129.1, 136.0, 166.2, 175.3

2,2-Bis(4-(1,5-bis(4-carbonylphenyl)-pentyl-3-oxy)phenyl)propane **8**

1,5-Bis(4-methoxycarbonylphenyl)-3-pentanol (7.2g, 20 mmol), 2,2-bis(4-hydroxyphenol)-propane (2g, 8.8 mmol), and triphenylphosphin (5.3g, 20 mmol) were dissolved in 100 mL THF and the solution was cooled to 0°C. Diethylazodicarboxylate (3.5g, 20 mmol), dissolved in 10 mL THF, was added slowly and the solution was stirred 24 h at room temperature. The solvent was removed and 50 mL methanol and 50 mL 1n aqueous NaOH solution were added. The solution was refluxed for 3 h, methanol was removed, the residue was filtered, diluted with 150 mL water, and extracted with 200 mL ethyl acetate. The pH of the aqueous phase was adjusted to 3 and the phase was extracted 3 x with 100 mL ethyl acetate. The combined organic phases were dried over Na_2SO_4. The solvent was removed in vacuum. Yield: 3g (40% th.)

^{1}H NMR (d$_6$-DMSO, δ in ppm): 1.58, 1.96, 2.75, 4.29, 6.79, 7.09, 7.27, 7.84

^{13}C NMR (d$_6$-DMSO, δ in ppm): 30.9, 30.7, 34.7, 41.0, 75.8, 115.1, 127.4, 128.3, 128.4, 129.3, 142.6, 147.0, 155.7, 167.0

Tri- and tetradendrons

Reaction with **6**

The monodendrons were dissolved in DMF, triethyl amine and **6** were added and the solution was stirred for 1-2 days. The solvent was removed and the residue was dissolved in ethyl acetate. The organic phase was extracted with aqueous $NaHCO_3$ and diluted HCl. Alternatively, the products were isolated by precipitation in methanol.

Reaction with **7**

Core **7**, **G2-Amine**, TBTU (O-(benzotriazol-1-yl)1,1,3,3,-tetramethyluroniumtetra-fluoroborate), and triethylamine were dissolved in DMF and stirred for 3 days at room temperature. The isolation of the product proceeded similarly as described for **6**.

Reaction with **8**

Core **8**, monodendron-Amine, TBTU (O-(benzotriazol-1-yl)1,1,3,3,-tetramethyluronium-tetrafluoroborate) or HATU (O-(7-azabenzotriazol-1-yl)-1,1,3,3,-tetramethyluronium-hexafluorophosphate (HATU), and triethylamine were dissolved in DMF and stirred for 3 days at room temperature. The isolation of the product proceeded similarly as described for **6**.

Acknowledgement

We would like to thank Dr. G. Pompe for DSC and thermal analysis measurements, and for helpful discussions. The financial support by *Deutsche Forschungsgemeinschaft* is gratefully acknowledged.

References

1) D. A. Tomalia, H. D. Durst, *Top. Curr. Chem.* **165**, 194 (1993); B. I. Voit, *Acta Polym.* **46**, 87 (1995); G. R. Newkome, C. N. Moorefield, F. Vögtle, *„Dendritic Molecules – Concept Syntheses Perspectives"*, VCH, Weinheim 1996; J. M. J. Fréchet, C. J. Hawker, Synthesis and Properties of Dendrimers and Hyperbranched Polymers in *Comprehensive Polymer Science*, Second Supplement, Eds.: S.L. Aggarwal, S. Russo, Elsevier Science Ltd., Oxford 1996, p. 71 and references therein.

2) G.R. Newkome, C.N. Moorefield, G.R. Baker, *Aldrichim. Acta* **25**, 31 (1992).

3) P.R. Ashton, D.W. Anderson, C.L. Brown, A.N. Shipway, J.F. Stoddart, M.S. Tolley, *Chem. Eur. J.* **4**, 781 (1998)

4) R.G. Denkewalter, J.F. Kolc, W.J. Lukasavage, *U.S.Pat. 4,410,688,* (1983).

5) S.M. Aharoni, C.R. Cresby III, E.K. Walsh, *Macromolecules* **15**, 1093 (1982).

6) P.M. Bayliff, W.J. Feast, D. Parker, *Polym. Bull.* **29**, 265 (1992).

7) S.C.E. Backson, P.M. Bayliff, W.J. Feast, A.M. Kenwright, D. Parker, R.W. Richards, *Macromol. Symp.* **77**, 1 (1994).

8) T.M. Miller, T.X. Neenan, *Chem. Mater.* **2**, 346 (1990).

9) K.E. Uhrich, S. Boegeman, J.M.J. Fréchet, S.R. Turner, *Polym. Bull.* **25**, 551 (1991).

10) M. Tang, C.T. Redemann, F.C. Szoka, *Bioconjugate Chem.* **7**, 703 (1996)

11) R. Duncan, N. Malik, N., *Proc. Int. Symp. Controlled Release Bioact. Mater.* **23**, 105 (1996).

12) Y.H. Kim, *J. Am. Chem. Soc.* **114**, 4947 (1992); Y.H. Kim, *Macromol. Symp.* **77**, 21 (1994).

13) H. R. Kricheldorf, G. Löhden, *J. Macromol. Sci., Pure Appl. Chem.* **A32**(11), 1915 (1995) and H. R. Kricheldorf, O.Bolender, T. Stukenbrock, *Macromol. Chem. Phys.* **198**, 2651 (1997)

14) S. Russo, A.Boulares, *Macromol. Symp.* **128**, 13 (1998)

15) M. Jikei, Y. Gang, C. S. Hyun, M.- A. Kakimoto, *Polym. Mater. Sci. Eng.* **77**, 200 (1997).

16) B.I. Voit, D. Wolf, *Tetrahedron* **53**, 15535 (1997)

17) A. R. Brenner, D. Schmaljohann, D. Wolf, B. I. Voit, *Macromol. Symp.* **122**, 1022 (1997)

18) S.J.E. Mulders, A.J. Brouwer, P.G.J. van der Meer, R.M.J. Liskamp, *Tetrahedron Lett.* **38**, 631 (1997)

19) O. Mitsunobu, *Synthesis* **1981**, 1.

20) W.W. Blount, J.R. Zoeller, US Patent 5 025 086 (1991); *Chem. Abstr.* **115**, 184133t (1991)

21) E. M. Fry, *J. Org. Chem.* **15**, 802 (1950)

22) N. Chau, S. Matsuda, Y. Iwakura, *Makromol. Chem.* **180**, 1435 (1979)

23) Ger. 1050540 (1959), Hoechst AG.,invs.: A. Jäger; *Chem. Abstr.* **55**, 5040 (1960)

24) H. L. Wehrmeister, *J. Org. Chem.* **28**, 2587 (1963) and *J. Org. Chem.* **28**, 2589 (1963)

25) P. A. Gunatillake, G. Odian, D. A. Tomalia, *Macromolecules* **20**, 2356 (1987)

26) Ger. 1062253 (1959), Hoechst AG., invs.: A. Jäger; *Chem. Abstr.* **55**, 13380 (1961)

[27] US 4613662, Ashland Oil, Inc., invs.: A. B. Goel; ; *Chem. Abstr.* **106**, 67870 (1987)
[28] C. Wörner, P. Müller, R. Mülhaupt, *Polym. Bull. (Berlin)* **34**, 301 (1995).
[29] T. Huber, F. Böhme, H. Komber, J. Kronek, J. Luston, D. Voigt, B. Voit, *Macromol. Chem. Phys.*, in print

Photoacid-generating Sulfonyloxymaleimide Polymers and Their Application to Photoimaging

Kwang-Duk Ahn* and Chan-Moon Chung†

Functional Polymer Laboratory, Biomaterials Research Center, Korea Institute of Science and Technology, P. O. Box 131, Cheongryang, Seoul 130-650, Korea

† Department of Chemistry, Yonsei University, Wonju, Kangwon-do 220-710

SUMMARY: Synthesis and properties of photoacid generating polymers based on four sulfonyloxymaleimides (RsOMI), *N*-(tosyloxy)maleimide (TsOMI), *N*-(methanesulfonyloxy)maleimide (MsOMI), *N*-(trifluoromethanesulfonyloxy)maleimide (TfOMI), and *N*-(10-camphorsulfonyloxy)maleimide (CsOMI) are described. The RsOMI polymers photochemically produce corresponding sulfonic acids (RsOH), TsOH, MsOH, TfOH and 10-camphorsulfonic acid (CSA) which can catalyze deprotection of the acid-labile t-BOC groups in the same polymer chains via a chemical amplification process. The photoacid generating polymers with concurrent acid-labile groups offer great potential for applications in functional transformation and photoimaging.

Introduction

Great deals of research efforts have been made for advancement of *N*-substituted functional maleimide monomers (R_fMI) and their polymers[1]. Functional maleimides offer a few unique features as follows: facile synthesis by introduction of a desired functionality into the imino nitrogen, relatively low reactivity in radical homopolymerization so as to handle and store without special cautions, and sufficiently high reactivity in copolymerization by electron-donor/acceptor (EDA) systems so as to obtain high molecular weight polymers with high conversions. Previously, we have reported on the *N*-substituted functional polymaleimides having acid-labile protecting groups represented by the *tert*-butyloxycarbonyl (t-BOC) group[2]. Those maleimide polymers have been applied to microlithography as chemical amplification resists in a deep UV (DUV) region. Polymaleimides containing trialkylsilyl moieties have been also studied as both the acid-labile polymers and photoresist materials[3,4].

In designing an advanced photoresist system based on the chemical amplification concept, it has to incorporate two basic components, namely, an acid-labile polymer and a photoacid generator (PAG) which should generate a strong acid by UV irradiation[5,6]. There has been reported numerous photoacid generating systems from the well-known onium salt PAGs to pure organic PAGs[7]. Nonionic sufonate PAGs have been extensively studied because they offer the advantages of better solubility and being free of metal ions compared to onium salt

 CCC 1022-1360/99/$ 17.50+.50/0

PAGs[8,9]. Among them, 2-nitrobenzyl sulfonates have received special attention as a useful PAG because of ease of synthesis, good solubility, and low UV absorption[8-11]. Most of the reported organic PAGs were low-molecular-weight compounds exemplified by 2-nitrobenzyl sulfonates and *N*-(tosyloxy)phthalimide which generate sulfonic acids by UV irradiation and they have been applied in photoimaging as efficient PAGs[9,12,13]. Polymeric PAGs are considered to be advantageous in terms of thermal stability and compatibility or mixing in the polymer matrix as well as nonvolatility in comparison with the corresponding small molecule PAGs. Only a few polymeric PAGs have been reported, including copolymers of 2-nitrobenzyl *p*-styrenesulfonate[14] and 9-fluorenylideneimino *p*-styrenesulfonate[15] which produce polymer-bound sulfonic acids upon DUV exposure. The copolymers having *p*-(*tert*-butyloxycarbonyloxy)styrene (t-BOCSt) were utilized as a single-component resist system without any externally added PAG.

In this regard, synthesis and properties of the photoacid generating polymers based on four kinds of sulfonyloxymaleimides (RsOMI), *i.e.*, *N*-(tosyloxy)maleimide (TsOMI), *N*-(methanesulfonyloxy)maleimide (MsOMI), *N*-(trifluoromethanesulfonyloxy)maleimide (TfOMI), and *N*-(camphorsulfonyloxy)maleimide (CsOMI) are described[16-20]. The RsOMI units in the polymers were designed to be the potential PAGs because phthalimido-sulfonates were utilized to generate sulfonic acids when irradiated with UV light[9,13]. In particular, the CsOMI polymers considered to have desired properties due to the bulky camphorsulfonate group both PAG itself and subsequently generated acid. The capability of photochemical acid generation and applicability of the RsOMI polymers as a single-component resist system were investigated.

Results and Discussion

Synthesis and Polymerization of Sulfonyloxymaleimides[16,18,20]

Four sulfonyloxymaleimides (RsOMI), TsOMI, MsOMI, TfOMI, and CsOMI have been prepared by a retro-Diels-Alder reaction through thermolysis of the corresponding RsOMI/furan adducts as described in Scheme 1. The synthetic results of the adducts and RsOMIs are listed in Table 1. The RsOMI monomers as an electron-acceptor possess high radical polymerizability in EDA copolymerization with electron-donating styrene monomers (XSt) such as styrene (St), *p*-methylstyrene (MeSt), *p*-hydroxystyrene (HOSt), *p*-acetoxy-

Scheme 1

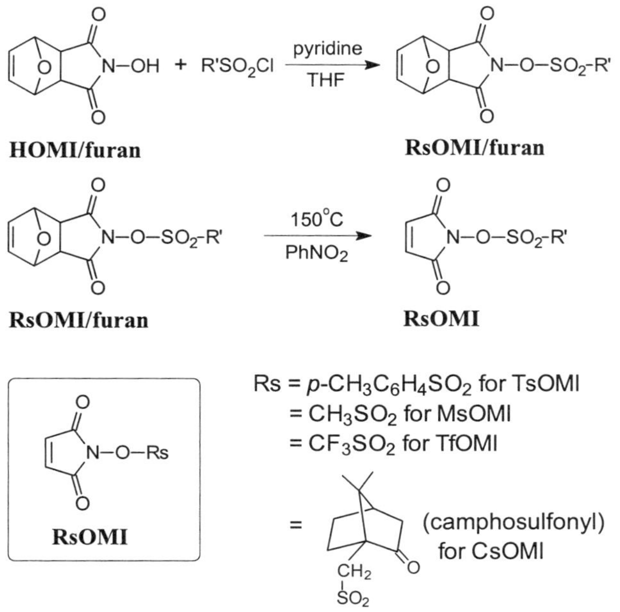

Table 1. Synthetic results of RsOMI/furan adducts and RsOMIs

Rs	RsOMI/furan		RsOMI		RsOMI
	mp (°C)	yield (%)	mp (°C)	Yield (%)	
$CH_3C_6H_4SO_2$	165	85	150	57	TsOMI
CF_3SO_2	116	91	56	75	TfOMI
CH_3SO_2	165	92	113	65	MsOMI
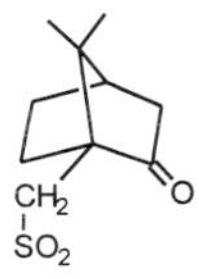	139	89	102	70	(+)CsOMI
	143	90	112	70	(±)CsOMI

styrene (AcOSt), and t-BOCSt. Terpolymers based on RsOMI and t-BOCSt were also prepared by use of maleimide (MI), *N*-(hydroxy)maleimide (HOMI), *N*-(*p*-hydroxyphenyl)-maleimide (HOPMI), or *N*-(*tert*-butyloxycarbonyloxy)maleimide (t-BOCMI) as the third maleimide monomer (YMI). The unsubstituted (unprotected) MI, HOMI, and HOPMI units were introduced into the terpolymers to improve photosensitivity and adhesion to substrates since their polymers are regarded as "partially deprotected" forms. The copolymers and terpolymers derived by EDA polymerizations of RsOMIs are shown in Scheme 2.

Copolymerizations of RsOMIs with styrene monomers were carried out with 1 to 1 molar feed ratio, and in terpolymerizations, RsOMI, YMI as maleimide (MI, HOMI, HOPMI, or t-BOCMI), and t-BOCSt were fed in the molar ratio of 1:1:2. The results of the radical copolymerizations were summarized in Table 2. The polymers were obtained in high yields over 85% in relatively short time usually within 3 hours. The polymers have inherent viscosity in the range of about 0.5 to 1.0 dL/g in dioxane (Table2). The molecular weights of

Table 2. Radical polymerizations of RsOMIs with styrenes (XSt)

Polymer[a]	AIBN[b] (mol%)	M/S[c] (g/ml)	Time (h)	Yield (%)	Inherent viscosity[d]
P(CsOMI/St)	2	0.54	3	88	0.73[e]
P(CsOMI/HOSt)	2	0.35	5	65	0.43
P(CsOMI/t-BOCSt)	2	0.55	2	89	0.73[f]
P(CsOMI/MI/HOSt)	2	0.50	5	84	0.70
P(TsOMI/St)	1	0.25	3	91	0.55
P(TsOMI/t-BOCSt)	1	0.50	2	86	0.81
P(TsOMI/MI/t-BOCSt)	1	0.40	3	93	0.94
P(TsOMI/t-BOCMI/t-BOCSt)	0.5	0.33	5	91	0.69
P(TfOMI/St)	1	0.33	2	80	0.90
P(TfOMI/t-BOCSt)	1	0.33	3	87	0.90
P(TfOMI/MI/t-BOCSt)	1	0.33	3	89	0.90
P(MsOMI/St)	2	0.50	1	85	0.81
P(MsOMI/t-BOCSt)	2	0.50	1	94	0.72

[a]All of the copolymers have an alternating structure. Copolymerizations and terpolymerizations were carried out in 1:1 and 1:1:2 molar feed ratio, respectively, at 55°C in dioxane. [b]Mole % of the initiators AIBN to the combined monomers. [c]Ratio of the total monomer weight to the volume of dioxane. [d]Inherent viscosity measured at a concentration of 0.20 g/dl in dioxane at 25°C. [e]Molecular weights measured by GPC in a mixture of chloroform and THF (1:1 by vol), $M_W = 5.72 \times 10^5$. [f]$M_W = 5.80 \times 10^5$.

the polymers were controlled by adjusting the concentrations of the radical initiator and the monomers. The composition of the copolymers P(RsOMI/XSt) was confirmed to have the same molar ratio as the monomer feed ratio by NMR spectra and elemental analysis. The obtained copolymers are considered to have an alternating composition of the maleimide and styrene units since it is well-known that the radical copolymerization of an electron-poor monomer (maleimides) and an electron-rich monomer (styrenes) gives an alternating structure[21-24].

Thermal Properties of Polymers

Thermal properties of P(RsOMI/XSt) were investigated by utilizing thermal analysis techniques, differential scanning calorimetry (DSC) and thermogravimetric analysis (TGA). The DSC thermograms of the RsOMI polymers having t-BOCMI and/or t-BOCSt units revealed endothermic events in the range of about 150 °C to 180 °C due to the deprotection of t-BOC groups. In TGA thermograms, the amounts of the mass loss during the deprotection were found to be the same as the theoretically calculated amounts based on liberation of carbon dioxide and isobutylene (i.e. Fig. 2 and 3). The deprotection leads to the conversion of t-BOCMI and t-BOCSt units into MI and HOSt units, respectively, which was confirmed by FT-IR and ^1H NMR spectral analyses. The deprotected polymers with MI and/or HOSt units showed high T_gs at above 200 °C.

The RsOMI polymers revealed extraordinary strong exothermic events beginning at about 220 °C for CSOMI, 225 °C for TfOMI, 260 °C for TsOMI, and 265 °C for MsOMI units in DSC analysis. The events are considered as cleavage of the corresponding sulfonyloxy groups from the polymers since the mass loss measured by the TGA thermograms nearly matches with the calculated amounts of the sulfonyloxy groups in each polymer. Immediately after the cleavage of sulfonyloxy groups the main-chains of the polymers eventually began to decompose.

Photochemical Acid Generation[16,18,20]

The photochemical generation of RsOH from P(RsOMI/XSt) was confirmed by isolation and identification of the acid. For example, the photogenerated camphorsulfonic acid (CSA) from P(CsOMI/HOSt) was isolated by solvent extraction from the UV-exposed polymer film and confirmed with ^1H NMR spectral analysis. It was also verified by photolysis of *N*-(10-camphorsulfonyloxy)succinimide as a model compound, which generated CSA and succinimide[20,25].

Scheme 2

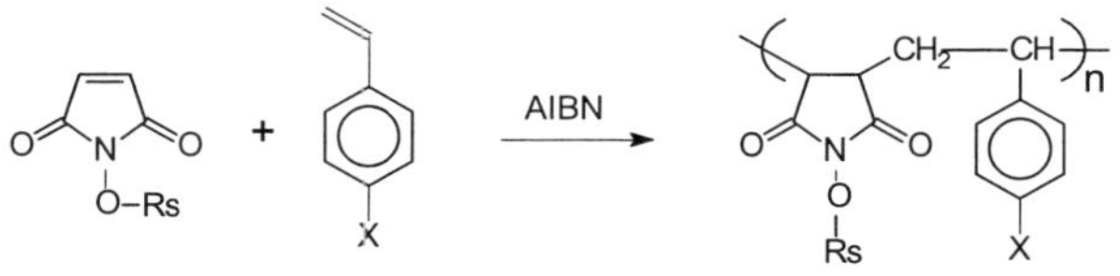

RsOMI **XSt** **P(RsMOMI/XSt)(1:1)**

(molar feed ratio =1:1)

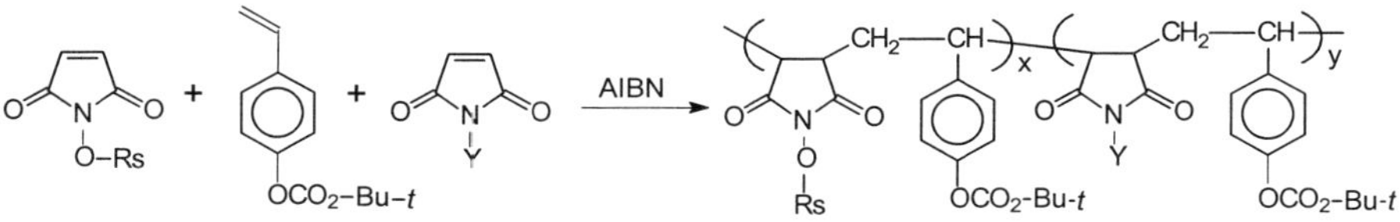

RsOMI **t-BOCSt** **YMI** **P(RsOMI/YMI/t-BOCSt)(1:1:2)**

(molar feed ratio =1:2:1)

Rs = Ts, CH_3SO_2, CF_3SO_2, camphorsulfonyl

X = H (St), Me (MeSt), OH (HOSt), OCO_2-Bu-*t* (t-BOCSt), OAc (AcOSt)

Y = H (MI), OH (HOMI) , C_6H_4OH-*p* (HOPMI) ,CO_2-Bu-*t* (t-BOCMI)

Scheme 3

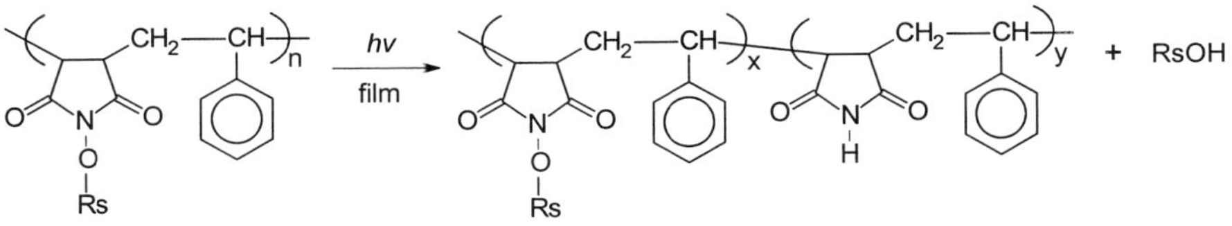

P(RsOMI/St) **P(RsOMI/MI/St)**

Rs = Ts, CH_3SC_2, CF_3SO_2, camphorsulfonyl

In photolysis of the P(RsOMI/XSt) films, not all of the RsOMI units yielded the corresponding sulfonic acid (RsOH) but only x mole fractions are converted into RsOH and MI units as described by the photolyzed polymer structure in Scheme 3. The amounts of RsOH generated by photolysis in film were determined by colorimetry utilizing a merocyanine dye or sodium salt of tetrabromophenol blue[16,25,26)]. The percent conversions of CsOMI units in P(CsOMI/XSt) at various UV exposure doses are shown in Figure 1. The degree of photoconversion of P(CsOMI/St) was slightly higher than that of P(TsOMI/St) or P(MsOMI/St). When hydroquinone (HQ) was added as a hydrogen donor, the photochemical generation of CSA in P(CsOMI/St) films was significantly enhanced. The tendency of sensitivity enhancement by the use of HQ was also observed in other P(RsOMI/St). A photosensitized electron-transfer mechanism has been postulated in the photoinduced N-O bond cleavage of *N*-acyloxyphthalimides with subsequent sulfonic acid generation in the same manner from dicarboximido (succinimido) sulfonates[12,13,27)]. Hydrogen donating additives like HQ in the photochemical reaction are known to enhance the capability of photochemical acid generation from the sulfonate PAGs[13)].

To provide the same enhancing effect as the HQ additive, HOSt units were incorporated into the CsOMI polymers by copolymerizing HOSt instead of St. The resulting phenolic polymer P(CsOMI/HOSt) revealed a remarkable increase in acid generation twice as much as that of P(CsOMI/St), and underwent photoconversion exceeding 70 mol % when irradiated with 1300 mJ/cm^2. P(CsOMI/HOSt) containing 5 wt % HQ exhibited the most effective acid generation among the CsOMI polymers with various combinations of comonomers and the HQ additive in this study.

Deprotection of t-BOC Groups by Sulfonic Acids Generated *in situ*[16-20)]

Upon exposure to 250-nm UV with 75 mJ/cm^2, the RsOMI polymers having acid-labile t-BOC groups, P(RsOMI/t-BOCSt) showed a distinctive endothermic event owing to the deprotection of t-BOC groups at much lower temperatures in their DSC thermograms than the deprotection temperatures (T_{dp}) of the unexposed polymers, implying the acidolytic deprotection by the photogenerated RsOH as depicted in Scheme 4. When the RsOMI polymers P(RsOMI/t-BOCSt)(1:1) were photolyzed, x mole fractions of RsOMI units generated RsOH to give P[$(RsOMI)_{1-x}$/$(MI)_x$/t-BOCSt], leaving t-BOC groups intact. In subsequent heating the acidolytic deprotection afforded fully deprotected polymers P[$(RsOMI)_{1-x}$/$(MI)_x$/HOSt]. The acidolytic deprotection of side-chain t-BOC groups catalyzed by the photogenerated RsOH was confirmed by TGA in which the observed amounts of mass

Scheme 4

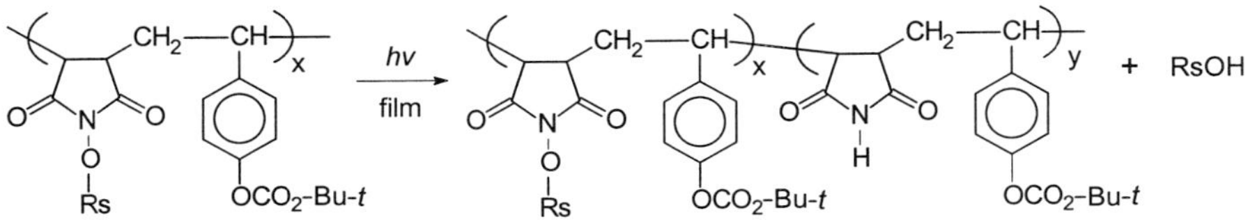

P(RₛOMI/t-BOCSt) **P(RsOMI/MI/t-BOCSt)**

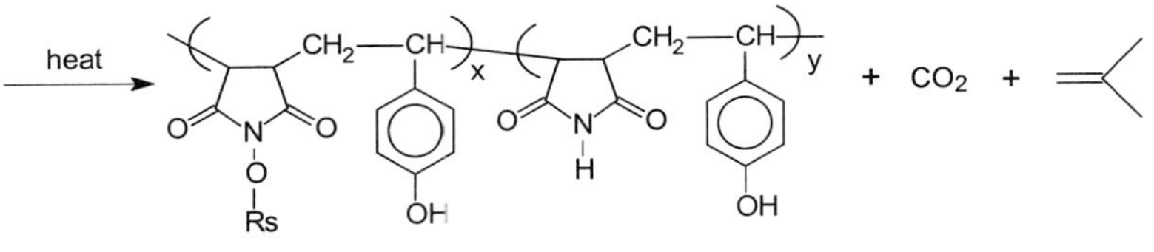

P(RsOMI/MI/HOSt)

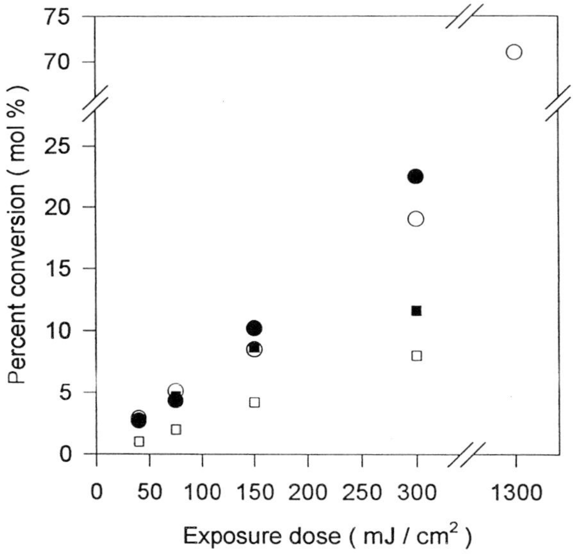

Fig. 1. Photochemical generation of camphorsulfonic acid from the CsOMI polymers by UV exposure in film: (○) P(CsOMI/HOSt); (●) P(CsOMI/HOSt) + 5% HQ; (□) P(CsOMI/St); (■) P(CsOMI/St) + 5% HQ.

loss caused by the deprotection at about the corresponding T_{dp} are equal to the theoretically calculated amounts.

In TGA thermograms as shown in Fig. 2, P(TsOMI/t-BOCMI/t-BOCSt) and P(TsOMI/MI/t-BOCST) reveal the equal amounts of the mass loss with 33% and 25%, respectively, at the different T_{dp}s before and after the UV exposure in the films[17)]. In Fig. 3 the TGA thermograms of P(CsOMI/t-BOCSt)(1:1) exhibit the same mass loss of 19% due to the complete deprotection of t-BOC groups by the photogenerated CSA after UV exposure[19)]. Upon exposure to 250nm-UV with 75mJ/cm^2, P(CsOMI/t-BOCSt) showed distinctive endothermic events in its DSC thermograms by the deprotection at the different T_{dp}s as depicted in Fig. 4[20)]. The acidolytic deprotection of P(CsOMI/t-BOCSt) was observed at 125°C quite lower than its original thermal deprotection temperature at 177°C without exposure. T_{dp} decreased further to 91°C when 5% HQ was added to the polymer as clearly shown in Fig. 4. The thermal deprotection behavior of P(RsOMI/t-BOCSt)(1:1) is much related to the nature of photogenerated acids (RsOH) so that the deprotection temperatures significantly decrease with the increase in the order of the acid strength CSA<TsOH<MsOH<TfOH. The deprotection temperatures of the t-BOC groups were observed by DSC and are summarized in Table 3[17,20)]. T_{dp}s decreased further when the polymers contained small amounts of HQ.

Table 3. Deprotection temperatures (T_{dp}) of t-BOC groups in the copolymers of RsOMI and t-BOCSt before and after exposure*

Polymer	Unexposed (°C)	Exposed (°C)	Exposed with HQ# (°C)
P(CSOMI/t-BOCSt) (1:1)	177	125	91
P(TsOMI/t-BOCSt) (1:1)	171	109	84
P(TsOMI/MI/t-BOCSt) (1:1:2)	171	116	93
P(TsOMI/t-BOCMI/t-BOCSt) (1:1:2)	171	120	94
P(MsOMI/t-BOCSt)(1:1)	166	110	60-90
P(TfOMI/t-BOCSt) (1:1)	150	87	79
P(TfOMI/MI/t-BOCSt) (1:1:2)	150	86	81

*Determined by DSC analysis before and after exposure to 250-nm UV with 75 mJ/cm^2 in film.
#Copolymers containing 10 % hydroquinone (HQ); 5 % HQ for terpolymers.

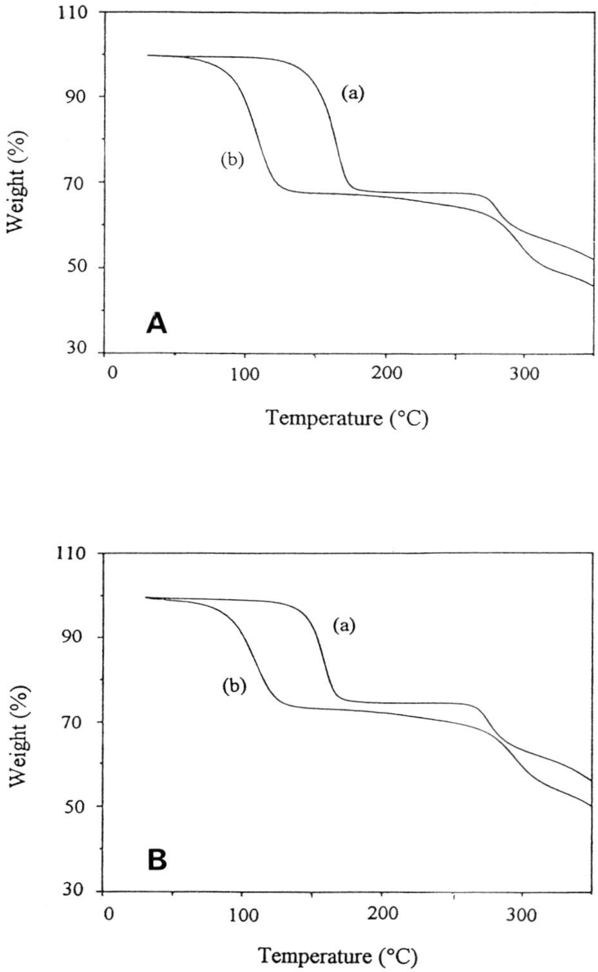

Fig. 2. TGA thermograms of (A) P(TsOMI/t-BOCMI/t-BOCSt) and (B) P(TsOMI/MI/ t-BOCSt) (a) before and (b) after exposure to 250-nm UV with 75 mJ/cm^2 for TsOH generation and deprotection in film.

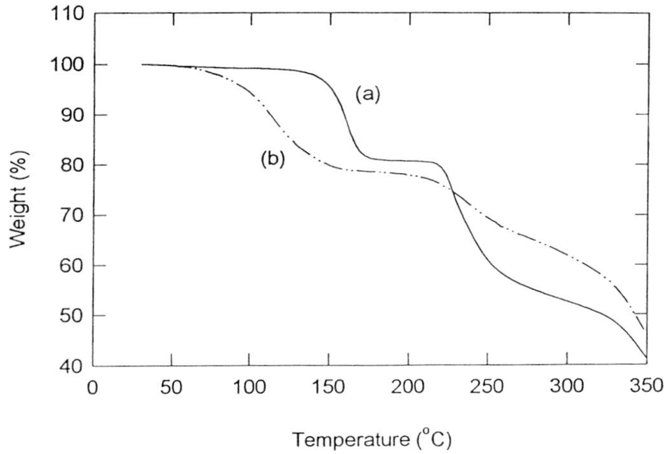

Fig. 3. TGA thermograms of P(CsOMI/t-BOCSt) (a) before and (b) after exposure to 250-nm UV with 100mJ/cm^2 for CSA generation and deprotection in film.

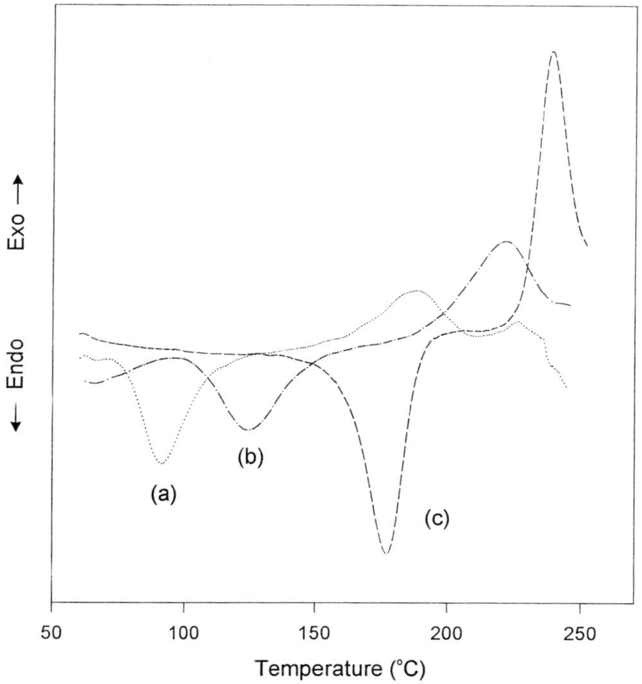

Fig. 4. DSC thermograms of P(CsOMI/t-BOCSt) after exposure to 250-nm UV with 75 mJ/cm^2: endothermic peaks appeared at (a) 91°C with 10% HQ, (b) 125 °C without HQ, (c) 177 °C without exposure.

Application as Single-Component Photoresists[17,19,28-30]

The copolymers and terpolymers of RsOMI have a good film-forming property and relatively low optical densities (OD) less than 0.3/μm at 250 nm except TsOMI terpolymers which have ODs of about 0.4/μm. All of the copolymers P(RsOMI/t-BOCSt) are soluble in common organic solvents such as chloroform, cyclohexanone, THF, and anisole, but insoluble in aqueous alkaline solutions such as 1.0 N NaOH and 2.38 wt % tetramethylammonium hydroxide (TMAH). However, they showed a significant change in solubility after DUV exposure and post-exposure bake (PEB) due to the large polarity change: the deprotected polymers are soluble in aqueous base developers.

The RsOMI polymers were employed as a single-component, chemically amplified photoresist without adding any PAG. The photoresist films were imagewise exposed by a contact mode using a 250-nm exposure system with doses of 50 to 100 mJ/cm^2 and then baked in order to induce acidolytic deprotection in the exposed areas. Negative- and positive-tone patterns were obtained by developing with organic and aqueous alkaline developers, respectively. We have extended the concept utilizing the RsOMI polymers having t-BOC groups to new single-component photoresists consisting of *N*-vinylpyrrolidone[28,29], and t-BOC-norbornene and *N*-(10-camphorsulfonyloxy)norborneneimide[30].

Photoinduced Functional Imaging

Selective transformation of chemical functionalities into the desired format is extremely worthwhile for making functional imaging onto thin polymer films[31-33]. Photoinduced reactions in the aforementioned RsOMI polymers are simple and versatile means to achieve the desired selectivity between the exposed and unexposed areas, leading to functional transformation in the defined (exposed) area. Upon incorporation of both the RsOMI and the acid-labile t-BOC group into a single reactive polymer chain, the resulting RsOMI polymers are willing to be transformed into new functional forms by virtue of photoacid generation and subsequent acid-catalyzed deprotection only in the exposed area. We have selectively generated such functional groups as free imide NH from t-BOCMI units, CO_2H from *t*-butyl ester moieties and phenolic OH from t-BOCSt units in the exposed polymer films.

Recently, we have prepared *N*-(10-camphosulfonyloxy)norborneneimide (CsNBI) and *N*-(tosyloxy)norborneneimide as functional norborneneimides (R_fNBI) and polymerized them to obtain new photoacid generating polymers[30]. These polymers having acid-labile protecting groups are again applicable as single-component photoresists in lithography, and as new functional polymers in photoimaging and functional imaging.

Acknowledgment. We thank all of the contributed coworkers and the financial support of MOST and KIST for the research project on Intelligent Polymers for Molecular Recognition.

References

[1] K.-D. Ahn, "*N-Protected Functional Polymaleimides*" in "*Polymeric Encyclopedia*", vol. **4**, J. C. Salamon, Ed.; CRC press, Inc., Boca Raton, Fl, USA, 1996, p. 2621
[2] C.-M. Chung, K.-D. Ahn, *Macromol. Symp.* **118**, 485 (1997)
[3] S.-T. Kim, J.-B. KIM, C.-M. Chung, K.-D. Ahn, *J, Appl. Polym. Sci.* **66**, 2507 (1997)
[4] K.-D. Ahn, C.-M. Chung, H.-S. Jo, J. M. Rhee, *Polym. Int.* **47**, 407 (1997)
[5] E. Reichmanis, F. M. Houlihan, O. Nalamasu, T. X. Neenan, *Chem. Mater.* **3**, 394 (1991)
[6] S. A. MacDonald, C. G. Willson, J. M. J. Fréchet, *Acc. Chem. Res.* **27**, 151 (1994)
[7] M. Shirai and M. Tsunooka, *Prog. Polym. Sci.* **21**, 1 (1996)
[8] J. Andraos, G. G. Barclay, D. R. Medrios, M. V. Baldovi, J. C. Scaiano, R. Sinta, *Chem. Mater.* **10** 1694 (1998)
[9] G. G. Barclay, D. R. Medeiros, R. F. Sinta, *Chem. Mater.* **7**, 1315 (1995)
[10] F. M. Houlihan, T. X. Neenan, E. Reichmanis, J. M. Kometani, T. Chin, *Chem. Mater.* **3**, 462 (1991)
[11] F. M. Houlihan, A. Shugard, R. Gooden, E. Reichmanis, *Macromolecules* **21**, 2001 (1988)
[12] J. I. G. Cadogan, A. G. Rowley, *J. Chem. Soc. Perkin I*, 1069 (1975)
[13] W. Brunsvold, R. Kwong, W. Montgomery, W. Moreau, H. Sachdev, K. Welsh, *Proc. SPIE* **1262**, 162 (1990)
[14] J. E. Hanson, E. Reichmanis, F. M. Houlihan, T. X. Neenan, *Chem. Mater.* **4**, 837 (1992)
[15] M. Shirai, H. Kinoshita, M. Tsunooka, *J. Polym. Sci., Polym. Chem.* **31**, 451 (1993)
[16] K.-D. Ahn, C.-M. Chung, D.-I. Koo, *Chem. Mater.* **6**, 1452 (1994)
[17] C.-M. Chung, D.-I. Koo, K.-D. Ahn, *J. Photopolym. Sci. Technol.* **7**, 473 (1994)
[18] K.-D. Ahn, J.-S. Koo, C.-M. Chung, *J. Polym. Sci., Polym. Chem.* **34**, 183 (1996)
[19] C.-M. Chung , K.-D. Ahn, *J. Photopolym. Sci. Technol.* **9**, 553 (1996)
[20] C.-M. Chung, K.-D. Ahn, *React. Funct. Polym.,* in press.
[21] H. K. Hall, Jr., A. B. Padias, *Acc. Chem. Res.* **23**, 3 (1990) and **30**, 332 (1997)
[22] S. R. Turner, C. C. Anderson, K. M. Kolterman, *J. Polym. Sci., Polym. Lett.* **27**, 253 (1989)
[23] K.-D. Ahn, D.-I. Koo, C. G. Willson, *Polymer* **36**, 2621 (1995)
[24] J. M. G. Cowie, Ed., *"Alternating Copolymers"*, Plenum Press, New York, USA, 1985
[25] M. Shirai, T. Yatsuo, M. Tsunooka, *J. Photopolym. Sci. Technol.* **9**, 273 (1996)
[26] G. Buhr, R. Dammel, C. R. Lindley, *ACS Proc. Polym. Mater. Sci. Eng.* **61**, 269 (1989)
[27] K. Okada, K. Okamoto, M. Oda, *J. Am. Chem. Soc.* **110**, 8736 (1988)
[28] S.-T. Kim, J.-B. Kim, J.-M. Kim, C.-M. Chung, K.-D. Ahn, *J. Photopolym. Sci. Technol.* **10**, 489 (1997)
[29] S.-T. Kim, J.-B. Kim, M.-H. Jung, K.-D. Ahn, *Polym. Bull.* **39,** 423 (1997)
[30] C.-W. Lee, J.-H. Shin, J.-H. Kang, J.-M. Kim, D.-H. Han, K.-D. Ahn, *J. Photopolym. Sci. Technol.* **11**, 405 (1998)
[31] A. M. Vekselman, C. Zhang, G. D. Darling, *Chem. Mater.* **9,** 1942 (1997)
[32] I. Willner, *Acc. Chem. Res.* **30**, 347 (1997)
[33] J. E. Beecher, G. H. McGall, M. J. Goldberg, *ACS Proc. Polym. Mater. Sci. Eng.* **77**, 394 (1997)

Complexation Behaviour of Polymers with Pendant Cyclodextrin Side Groups

Jörg Adams*, Markus Hollas, Myung-Ae Chung

Institut für Physikalische Chemie, TU Clausthal, 38678 Clausthal-Zellerfeld, Germany

Joon Woo Park[1], Ji Hyun Hong[1], Kwanghee Koh Park[2]

[1] Department of Chemistry, Ewha Womans University, Seoul 120-750, Korea

[2] Department of Chemistry, Chungnam National University, Taejeon 305-764, Korea

SUMMARY: The influence of the chain conformation on the formation of polymeric supramolecular complexes as well as the influence of the complexation on the conformation of the polymer chain has been studied. The complexation of pyrene into the cavity of β-cyclodextrin (β-CD) was investigated in aqueous solutions of β-CD substituted poly(allylamine) (PAA) under variation of external parameters, i.e. temperature, pH, ionic strength and addition of urea. The observed changes of the complexation constant K for the formation of the 2:1 β-CD/pyrene complex can be explained by a change of the chain flexibility which leads to a variation of the mean distance between neighbouring β-CD-moieties along the polymer chain. The intra-chain association of the decyl group with β-CD in PAA with co-pendant decyl and β-CD is disrupted by the addition of 1-adamantanamine HCl resulting in a more extended structure of the polymer. The β-CD moiety in PAA-CD shows one order of magnitude greater affinity to 2-(p-toluidyl)naphthalene-6-sulfonate than the native β-CD and the affinity increases further by the presence of decyl side groups.

Introduction

Covalent bonding of supramolecular functional units to a polymer backbone may lead to a large change in the physicochemical properties of the polymer as well as the functional units[1-3]. If the polymer and the functional units co-operate in a synergistic manner, new, highly selective materials are to be expected for analytical and catalytic processes. In this way it might be possible to mimic the high efficiency of biopolymers.

 CCC 1022-1360/99/$ 17.50+.50/0

Because of their good water solubility and the hydrophobic cavity, cyclodextrins (CDs) are ideal functional units for the complexation of organic compounds in aqueous solutions. They also can easily be attached to the polymer backbone of water soluble polyelectrolytes.

In this paper we want to discuss two different approaches to study the influence of the polymer on the complexation behaviour. The first part deals with the control of the complexation by controlling the chain conformation and the second part focuses on the control of the chain conformation by displacing the guest of a polymeric, intramolecular complex by a low molecular weight species.

For the concept of the first part one has to consider that the stoichiometry of a CD host-guest complex depends on both the ring size of the CD and the size of the guest molecule. In case of a 2:1 complex between two polymer-bound CD units and one low molecular weight guest molecule a strong impact of the polymer chain on the complexation is to be expected, because of the necessary co-operative process.

The fluorophor pyrene is suited as guest molecule because of its ability to form such 2:1 complex structures with β-CD and furthermore pyrene allows to detect the complexation process via its sensitivity towards micropolarity[4)]. In a previous paper[5)], we showed that the 2:1 complex between pyrene and β-CD attached to poly(allylamine) (PAA) in aqueous solutions has a chelate-type structure (i.e. no 1:1 intermediate) and is strongly controlled by the distance between neighbouring CD groups along the polymer chain.

The chain conformation of PAA is controlled by H-bonds and electrostatic interaction between neighbouring amino- and ammonium groups. Therefore, the desired variation of the conformation is possible by adjusting pH, ionic strength and temperature of the polymer solution. Different energetic situations of the chain should lead to a change in the entropy contributions to the driving force of the complexation.

Also, the interaction between the polymer-bound CDs and a guest can affect the properties of the polymer chain. This is the object of the second part of this paper. Therefore we prepared PAAs with co-pendant alkyl and β-CD side groups. Intra-chain association between the side groups is expected in these polymers. This should change the conformation and thus the substrate binding properties of the host polymers.

Experimental

Poly(allylamine) with pendant β-CD side groups (degree of substitution DS = 2%, 3.6%, 9% and 11.3%) was synthesized by a procedure published in ref. 5. Decyl groups were linked to PAA and some of the PAA-CD polymers by reacting the appropriate amounts of 1-bromodecane with the polymers in methanol / water (4:1 V/V) media. After evaporating the solvent, the solid residue was washed with n-hexane to remove the unreacted 1-bromodecane and then dissolved in water and recovered by freeze-drying. The DS with CD and decyl groups were calculated from NMR integration data. Corrected fluorescence spectra of the pyrene- or 2-(*p*-toluidyl)-napthalene-6-sulfonate (TNS)- containing aqueous solutions were taken with a SPEX FLUOROLOG 2 or a HITACHI F3010 spectrofluorimeter. For the experiments with aqueous pyrene solutions, the sample preparation and the automatic titration experiments with the electronic dispensing system EDOS 5221 (EPPENDORF) were carried out according to ref. 5. The pyrene concentration was always set to $2 \cdot 10^{-7}$ mol/L. The pH of the solutions were adjusted with HCl and measured with an electronic pH-meter. β-CD was obtained from WACKER CHEMIE GmbH, all other chemicals were obtained from ALDRICH and were used without further purification. Viscosity measurements were made with an *Ubbelohde* capillary viscometer in a tempered water basin. The equilibrium binding constants $K_{0\rightarrow1}$, $K_{1\rightarrow2}$ and K_{Ch} (see Results and Discussion) were calculated from the first (I_1, λ = 373 nm) and third vibronic transition (I_3, λ = 384 nm) of the pyrene emission spectrum (λ_{Ex} = 335 nm) according to *Xu et al.*[1]. Changes in the micropolarity in the vicinity of the pyrene probe molecules are expressed via the intensity ratio $R = I_1/I_3$. A high value of this quantity corresponds to a high polarity and a low value to a nonpolar local environment[4].

Equilibrium dialysis was carried out with a membrane of MW cut-off 500 for at least 7 days. Concentrations of TNS were determined from fluorescence intensity in β-CD solutions. The results of inside→out and outside→in diffusion were averaged. The equilibrium constants $K_{0\rightarrow1}$ and $K_{1\rightarrow2}$ for the subsequent complexation of TNS by one and two β-CDs were calculated by fitting the TNS intensity profile according to *Martel et al.*[2].

Results and Discussion

Pyrene Complexation Behaviour of Poly(allylamine) with β-CD Side Groups.

As shown previously[5)] the 2:1-complexation mechanism of pyrene in PAA-CD depends strongly on the degree of substitution (DS) of the PAA with β-CD: at low DS (< 5%) the complexation is a 2-step process, equivalent to native β-CD, i.e. with an 1:1 intermediate and the associated equilibrium constants $K_{0\rightarrow 1}$ and $K_{1\rightarrow 2}$ (Fig. 1). However, at higher DS the binding of pyrene can only be explained by a chelate-type complexation, i.e. a one-step process with K_{Ch} as the overall complexation constant. This is due to the smaller distance between the CDs along the polymer chain.

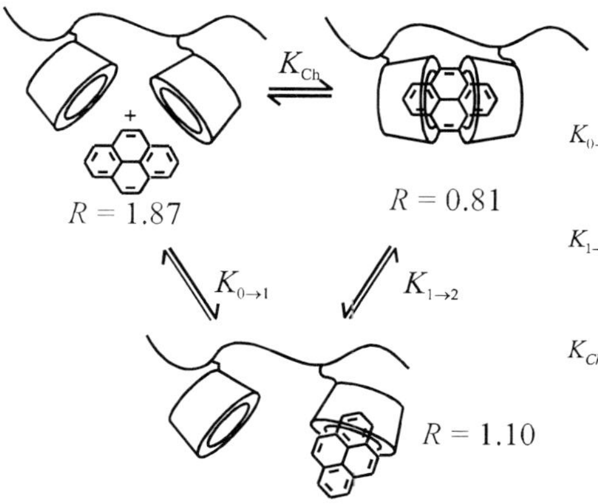

$$K_{0\rightarrow 1} = \frac{[\mathrm{Py-CD}]}{[\mathrm{Py}]\cdot[\mathrm{CD}]}$$

$$K_{1\rightarrow 2} = \frac{[\mathrm{CD-Py-CD}]}{[\mathrm{Py-CD}]\cdot[\mathrm{CD}]}$$

$$K_{Ch} = \frac{[\mathrm{CD-Py-CD}]}{[\mathrm{Py}]\cdot[\mathrm{CD-CD}]} = \frac{[\mathrm{CD-Py-CD}]}{[\mathrm{Py}]\cdot 0.5\cdot[\mathrm{CD}]}$$

Fig. 1: Schematic representation of the pyrene complexation by PAA-CD. The given *R* values of the individual species are calculated from the fluorescence intensities I_1 and I_3 according to ref. 1.

As mentioned above, a change in the mean distance between the CD units and thus the equilibrium constant is also expected by varying the temperature. Fig. 2 shows the ln *K* versus 1/T plot of the binding constant for the polymer with DS = 9% and a degree of polymerization P_{n} of 150-200.

In comparison with the data for native β-CD[5)] (also shown in Fig. 2), *K* of the polymer decreases much stronger with increasing temperature. The data analysis also revealed a change in the complexation mechanism from chelate-type complexation below 35 °C to subsequent 1:1 and 2:1 complexation above this temperature.

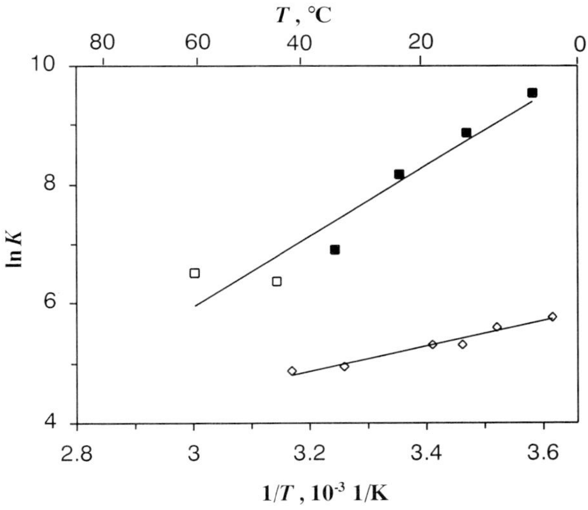

Fig 2: Temperature dependence of the equilibrium constant K for PAA-CD (DS = 9%) (□, ■) and β-cyclodextrin (◊). Closed symbols: $K = K_{Ch}$, open symbols: $K = (K_{0\rightarrow 1} \cdot K_{1\rightarrow 2})^{1/2}$; —: regression lines.

The thermodynamic data calculated from Fig. 2 are ΔH = -34.4 kJ/mol, and ΔS = -29.2 J/(mol·K) for β-CD and ΔH = -49 kJ/mol, ΔS = -95 J/(mol·K) for PAA-CD. This shows that both systems are enthalpy driven with a negative reaction entropy. In case of the CD polymer a smaller entropy loss compared to native β-CD was expected due to a restricted mobility of the CD units. The observed higher entropy loss is attributed to the temperature dependence of the polymer chain flexibility. Viscosity measurements of aqueous solutions of the PAA-CD exhibited a significant coil expansion with increasing temperature. A decreasing fraction of lipophile interacting methylene- and methine groups is the reason for this coil expansion, which leads to increasing distances between neighbouring CD units. As a consequence, the complexation constant decreases much stronger with increasing temperature compared to the native β-CD.

The lower ΔH value of the CD polymer is not a result of a direct interaction of the pyrene guest with the polymer backbone in the investigated concentration range, as proved by titration experiments with pure PAA[5]. Hence, stronger interactions between the polymer backbone and CD moieties in the complexed state have to be assumed.

The pH shows a significant influence on the reduced viscosity of aqueous solutions of PAA-CD (DS = 9%) (Fig. 3). At pH = 7 the typical viscosity behaviour of polyelectrolytes can be observed. On first glance the decrease of the viscosity in acidic solutions contradicts to this behaviour, but can be explained with the strong control of the conformation by H-bonds along the polymer chain[2)]. At pH = 7 the polymer chain has the most extended conformation because of a high number of stiffening H-bonds between amino- and ammonium groups. Increasing the degree of ionization leads to a decrease of the number of this H-bonds and consequently to smaller coil dimensions in spite of the electrostatic repulsion between the ammonium groups. On the other hand, alkaline solutions of CD polymers lead to a coil collapse because of a lack of the stabilizing electrostatic interactions. The observed strong viscosity decrease in salt solutions can be explained by partial screening of electrostatic interactions and breaking of H-bonds because of counter ion condensation.

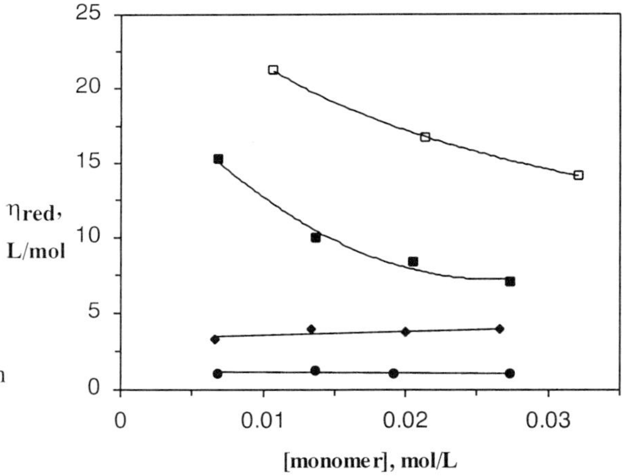

Fig. 3: Reduced viscosity of aqueous solutions of poly(allylamine), P_n = 150-200, at pH = 7 (□) and of PAA-CD (DS = 9%) at pH = 7 (■), pH = 0.5 (◆) and pH = 7, with [NaCl] = 3.5 mol/L (●); T = 295.1K. Polymer concentration is given in terms of allyl monomeric units.

The intensity ratios $R = I_1/I_3$ resulting from titration experiments in neutral, acidic or salt-containing solutions of PAA-CD (DS = 9%) are shown in Fig. 4. In acidic and neutral solutions pyrene is complexed in the same concentration range but the final R values are significantly higher in acidic solutions. This cannot be attributed only to a higher solvent polarity but rather indicates a change in complexation behaviour in favour of the 1:1 complex.

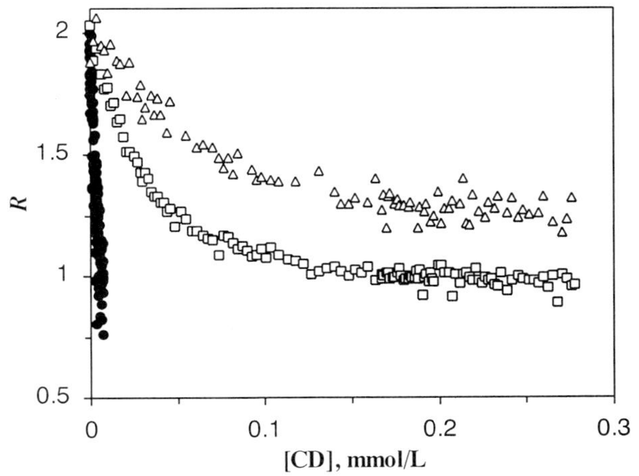

Fig. 4: Micropolarity $R = I_1/I_3$ versus CD concentration in aqueous solutions of PAA-CD (DS = 9%) at pH = 7 (□), pH = 1 (Δ) and pH = 7, [NaCl] = 3.5 mol/L (●); T = 295.1K.

The hindered formation of chelate-type structures in acid solutions might be explained with a folding of the polymer backbone around the CDs as already postulated by *Martel et al.*[2] for poly(vinylamine) in the presence of native β-CD. The chain folding is the result of numerous H-bonds between OH- groups of the CDs and ammonium groups of the polymer backbone. Furthermore, enthalpic interactions between complexed pyrene and the polymer backbone have to be assumed.

In the presence of salt the complexation of pyrene is shifted about two orders of magnitude to smaller CD concentrations (Fig. 4). This is a hint for the higher dominance of the chelate complex and a high equilibrium constant in comparison to neutral salt-free polymer solutions. Reason for this behaviour is an additional entropic contribution of the chain as a consequence of the reduced distance between neighbouring CDs. K_{Ch} could not be calculated for these solutions, because the pyrene concentration ([Py] = $2 \cdot 10^{-7}$ mol/L) is no longer negligibly small in comparison to that of CD (see ref. 1).

Similar to biopolymers the conformation may even be affected by denaturing agents[6] like urea. If H-bonds control the polymer structure in PAA-CD, the distance between adjacent CD moieties should therefore be controllable by the addition of urea. However, the interactions of urea with organic molecules in the presence of CDs are not fully enlightened up to now[7]. Interactions of urea with both the polymer and pyrene are possible in PAA-CD.

The hindered formation of chelate structures in the presence of urea is indicated by a significant shift of the decrease of R on the concentration scale (Fig. 5). Reason for this is an energetic stabilization of free pyrene in combination with a coil expansion attributed to decreasing hydrophobic interactions and a break up of H-bonds. K_{Ch} could not be determined for these solutions.

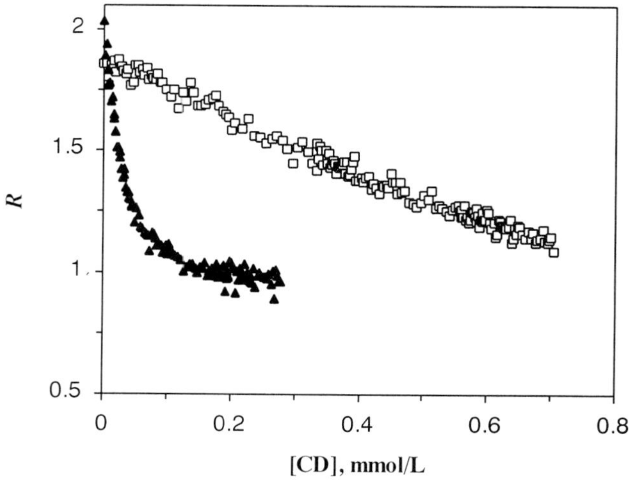

Fig. 5: Pyrene fluorescence intensity ratio R as a function of CD concentration for PAA-CD (DS = 9%) in the absence (▲) and presence (□) of urea (c = 5 mol/L); [Py] = $2\cdot10^{-7}$ mol/L, T = 293.1 K.

Viscosity and TNS Binding Properties of Poly(allylamine) with Co-pendant Decyl and β-Cyclodextrin Side Groups.

Alkyl chains show high binding affinity towards the β-CD cavity[8]. Thus the intra-chain association of alkyl with β-CD is expected in PAA with co-pendant alkyl and β-CD groups (PAA-alkyl-CD), resulting in a compact structure of the polymer. The addition of 1-adamantanamine HCl, of which the binding constant with native β-CD was reported as 18,000 L/mol[9], has little effect on the viscosity of PAA solution, but increases the viscosity of PAA-decyl-CD solution (Fig. 6).

Also, the intrinsic viscosity of PAA-decyl-CD (DS of decyl 6%; DS of CD 3.6%) increased from 1.05 dL/g in the absence of 1-adamantanamine HCl to 1.08 dL/g in the presence of 6 mM 1-adamantanamine HCl at pH 3.1, ionic strength 0.04 mol/L and 25°C (data not shown). This clearly indicates that PAA-decyl-CD molecules indeed adopt a compact conformation by intra-chain association of decyl group with β-CD, but the association is disrupted by the inclusion of 1-adamantanamine in the β-CD moiety (Fig. 7).

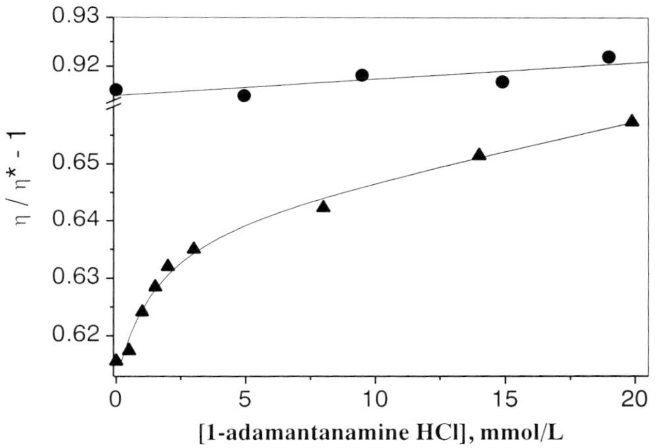

Fig. 6: Dependence of the specific viscosity of 5.0 g/L PAA (●) and PAA-decyl-CD (▲) on the concentration of 1-adamantanamine HCl at 25 °C, pH 3.1 and ionic strength $I = 0.04$ mol/L. The number of monomer units of PAA is 530-640, DS of decyl is 6%, and that of β-CD is 3.6%.

Another evidence for intra-chain association is the shape of the viscosity titration curve. The concentration of β-CD moiety in the solution used in Fig. 6 is 1.26 mmol/L. If the β-CD is not associated with the decyl group and the binding affinity for 1-adamantanamine is the same as that of free β-CD, the titration curve should show a saturation behaviour above about 1.5 mmol/L of 1-adamantanamine. In fact, the curve exhibits a rather heterogeneous binding and the average binding constant is less than 1000 L/mol. This is in good agreement with our expectation that the β-CD cavity is occupied with a decyl group and inclusion of 1-adamantanamine requires the release of the decyl unit.

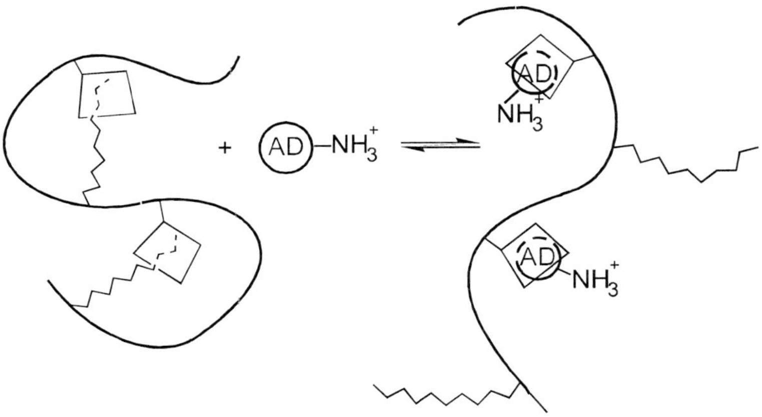

Fig. 7: Schematic representation of association of 1-adamantanamine with PAA-decyl-CD and structural change of the polymer.

TNS is virtually non-fluorescent in aqueous media. However, the addition of PAA or its derivatives and β-CD to aqueous TNS solutions enhances the fluorescence, reflecting transfer of the dye to apolar polymer domain. Fig. 8 shows the dependence of TNS fluorescence intensity on the concentration of polymers and β-CD. Analysis of the TNS fluorescence intensity profile in terms of 1:1 binding[2)] for β-CD and CD polymers did not give reasonable fitting. However, the fitting of TNS binding to β-CD to a scheme of 1:1 and 2:1 successive binding (Eq. 1) gave the $K_{0\to1}$ and $K_{1\to2}$ values of 3600 and 14 L/mol, respectively, and the fluorescence intensity ratio of the complex to the free TNS were found to be 21 for 1:1 complex and 173 for 2:1 (β-CD:TNS) complex. The binding constants are in reasonable agreement with the report of Kondo et al.[10)] and Jobe and coworkers.[9)]

$$I = \frac{I_{TNS} + I_{CD-TNS} \cdot K_{0\to1} \cdot [\mathrm{CD}] + I_{CD-TNS-CD} \cdot K_{0\to1} \cdot K_{1\to2} \cdot [\mathrm{CD}]^2}{1 + K_{0\to1} \cdot [\mathrm{CD}] + K_{0\to1} \cdot K_{1\to2} \cdot [\mathrm{CD}]^2} \quad (1)$$

I_{TNS}, $I_{\mathrm{CD\text{-}TNS}}$, and $I_{\mathrm{CD\text{-}TNS\text{-}CD}}$: fluorescence intensities of the free TNS, the 1:1 and 2:1 complex respectively.

Though the results of fluorescence titration (Fig. 8) do not give quantitative information on the TNS binding with the polymers, it is shown that the binding affinity of TNS to the polymers and the fluorescent quantum yield of the polymer-bound TNS is highly dependent on the nature of side groups and degree of substitution.

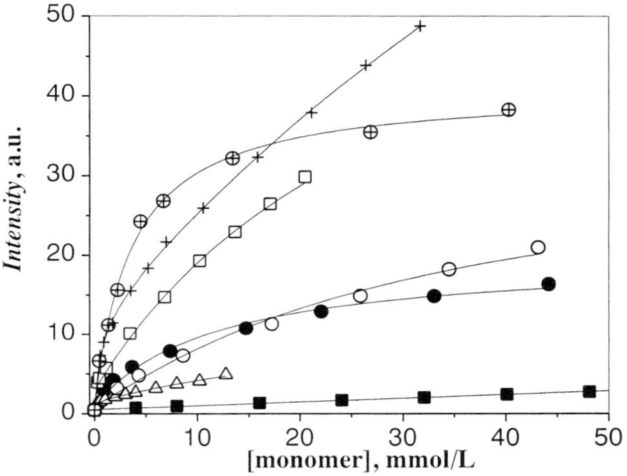

Fig. 8: Dependence of fluorescence intensity of $1.6 \cdot 10^{-6}$ mol/L TNS solutions at 438 nm on the concentration of PAA (■), β-CD(Δ), PAA-decyl (DS 6%) (●), PAA-CD (DS 2%) (O), PAA-CD (DS 11.3%) (□), PAA-decyl (DS 13%) (⊕), and PAA-decyl(DS 6%)-CD(DS 3.6%) (+) at 25 °C. The concentration of polymer is given in terms of allyl monomeric units.

From the initial slope of the variation of the fluorescence intensity on the concentration of polymers, it is estimated that the order of binding affinity of TNS to the polymers is

PAA < β-CD < PAA-CD(DS 2%) < PAA-decyl (DS 6%) < PAA-CD (DS 11.3%)
< PAA-decyl(DS 6%)-CD(DS 3.6%) < PAA-decyl (DS 13%).

Fig. 8 also indicates that the fluorescence quantum yield of bound TNS is much higher for the polymer with CD and / or decyl side groups than that for the native β-CD or unsubstituted PAA.

To obtain a quantitative information on the binding affinity of TNS with the polymer, we carried out equilibrium dialysis experiments. The results are summarized in Tab. 1: *K* refers to the apparent binding constant with the allyl unit of the polymer and *K*' to the binding constant per amine, decyl or CD functionality.

Table 1. Results of equilibrium dialysis of TNS in PAA and its derivative solutions ($pH = 3.0$, $I = 0.04$ mol/L)

polymer	$[P]_{tot}$	$[TNS]_{bound}$	$[TNS]_{free}$	K^a	K'
	10^{-2} mol/L	10^{-5} mol/L	10^{-7} mol/L	L/mol	L/mol
PAA	5.62	0.15	5.8	44	44
	2.81	1.30	60.6	80	80
PAA-decyl (DS 6%)	4.74	5.05	5.8	1800	30000[b]
	2.50	44.3	60.6	3000	50000[b]
PAA-decyl (DS 13%)	2.29	119	60.6	9000	70000[b]
PAA-CD (DS 2%)	4.93	2.15	5.8	750	40000[c]
PAA-CD (DS 11.3%)	1.22	29.2	60.6	4000	40000[c]
PAA-decyl-CD (DS 6%, 3.6%)	3.56	6.0	5.8	2900	
β-CD					3600[d], <16[e]

[a] Apparent 1:1 complexation constant of TNS with monomeric unit of the polymer
[b] Assumed 1:1 binding of TNS with decyl group of PAA-decyl.
[c] Assumed 1:1 binding of TNS with CD moiety of PAA-CD.
[d] $K_{0\to 1}$ of β-CD / TNS binding.
[e] $K_{1\to 2}$ of β-CD / TNS binding.

The binding constant of TNS with PAA-CD is one order of magnitude greater than that with the native β-CD. *Martel et al.*[2)] reported the binding constant of TNS with poly(vinylamine) at pH 9.5 as $0.3\text{-}2\cdot 10^4$ L/mol, which is significantly less than the value obtained here, presumably due to the absence of electrostatic effect in their systems.

The high binding affinity of TNS to CD moiety of PAA-CD seems to be the consequence of combined effects of electrostatic interaction with an ammonium group and hydrophobic interaction with the polymer backbone as well as the CD cavity. The agreement of the binding constant of the PAA-CD complex with that of the product of the binding constants of unsubstituted PAA and native β-CD complexes supports this.

The fraction of TNS bound to a polymer (denoted as P) becomes $K[P]/(1+K[P])$ for 1:1 binding. This relationship predicts that the concentration of polymers required for more than 90% binding is 200 mmol/L for PAA, 5 mmol/L for PAA-decyl (DS 6%), 12 mmol/L for PAA-CD (DS 2%), and 3.2 mmol/L for PAA-decyl(DS 6%)-CD(DS 3.6%). However, the fluorescence intensity *vs* concentration profile (Fig. 8) exhibits a large enhancement of fluorescence intensity by further increase of the polymer concentration, especially for the polymers with CD side groups. This implies a change in the microenvironment of the dye bound to the polymers. Two groups concluded that TNS forms 2:1 CD complexes with PAA-CD as pyrene complex (*vide infra*)[11]. However, this was rebutted by *Martel et al.*[2] on the basis of the dependence of the binding constant on DS and structural point of the polymer chain. Though the curves in Fig. 8 are well fitted to a scheme of 1:1 binding for the polymers without CD groups, the data of PAA-CD are fitted better to a model of successive 1:1 and 2:1 binding. This indicates that the 1:1 TNS/CD complex binds with another CD group to form a 2:1 complex in the polymer. In this case, the high stability of the 1:1 complex can be explained by the involvement of electrostatic interaction with ammonium groups. As with the native β-CD complexes, the 2:1 complexes exhibit much higher fluorescence quantum yield than the 1:1 complexes.

Conclusion

Complexation behaviour of a cyclodextrin polymer can be controlled by varying entropy contributions of the chain situation. Control parameters are temperature, pH, salt and urea concentration. Some of these parameters lead to additional reaction channels for pyrene. In these cases, qualitative comparison of complexation entropy with that in neutral, salt-free polymer solutions is difficult. Poly(allylamine) with co-pendant decyl and β-CD side groups adapts compact structure *via* intra-chain association of the decyl group with β-CD. The association is disrupted by a guest of high binding affinity to β-CD.

Derivatization of PAA with decyl and β-CD groups results in high affinity to TNS and large fluorescence intensity of the polymer bound TNS. The environment of TNS in the domain of PAA-decyl depends little on the dye / polymer ratio, but stoichiometry of CD/TNS complexes with PAA-CD changes to 2:1 from 1:1 as the concentration of polymer is higher.

Acknowledgement

This work was supported by a German-Korean joint research program (DFG Ad 99/2-2 and KOSEF 965-0300-001-2). We are thankful to WACKER CHEMIE, Munich, for supply of β-Cyclodextrin.

References

1) W. Xu, J. N. Demas, D. A. DeGraff, M. Whaley, *J. Phys. Chem.* **97**, 6546 (1993)

2) B. Martel, Y. Leckchiri, A. Pollet, M. Morcellet, *Eur. Polym. J.*, **31**, 1083 (1995)

3) M. Weickenmeier, G. Wenz, J. Huff, *Macromol. Chem., Rapid Commun.*, **18**, 1117 (1997)

4) D. Dong, M.A. Winnik, *Can. J. Chem.* **62**, 2560 (1984)

5) M. Hollas, M.-A. Chung, J. Adams, *J. Phys. Chem. B*, **102**, 2947 (1998)

6) A. L.- Lehninger, *Principles of Biochemistry*, Worth Publishers, Inc., New York 1982

7) L. A. Godinez, L. Schwartz, C. M. Criss, A. E. Kaifer, *J. Phys. Chem. B*, **101**, 3376 (1997)

8) (a) J. W. Park and H. J. Song, *J. Phys. Chem.* **93**, 6454 (1989)
(b) J. W. Park and K. H. Park, *J. Inclu. Phenom.,* **17**, 277 (1994)

9) D. J. Jobe, R. E. Verrall, E. Juquera, and E. Aicart, *J. Phys. Chem.*, **97**, 1243 (1993)

10) H. Kondo, H. Nakatan, and K. Hiromi, *J. Biochem.*, **79**, 393 (1976)

11) (a) T. Seo, T. Kajihara, K. Miwa, and T. Iijima, *Makromol. Chem.*, **192**, 2537 (1991)
(b) A. Harata, M. Furue, and S. Nozakura, *Macromolecules*, **10**, 676 (1977)

Macromol. Symp. **142,** *173–183 (1999)*

A New Photobase Generator Containing Oxime-urethane Group and Its Application

Kyu Ho Chae* and Du-Jeon Jang

*Department of Polymer Engineering, Chonnam National University, Kwangju 500-757, Korea

Department of Chemistry and Research Institute of Molecular Sciences, Seoul National University, Seoul 151-742, Korea

SUMMARY: Photochemical base formation from oxime-urethane derivatives was observed via pH changes, GC and HPLC analysis, and laser flash photolysis study. Photolysis of these derivatives results in the formation of amines which induce cross-linking of poly(glycidyl methacrylate) (PGMA) upon heating. Photocross-linking reaction of the copolymers, containing epoxy and oxime-urethane groups, were studied by measuring the insoluble fraction of copolymer films under various reaction conditions. Thermal cross-linking reaction of the copolymer was catalysed by the photogenerated pendant amines. The copolymer containing oxime-urethane groups which photogenerate aromatic amines can be utilized as an image recording material.

Introduction

Photoinitiators can be classified as radical or ionic photoinitiators based on their reaction mechanism. Although most of the industrially used photoinitiators are radical initiators, ionic ones have also become increasingly important in microlithography, photoresists, and coatings application. The advantages of ionic photoinitiators over radical ones are their ability to introduce ring opening polymerization, and lack of volume contraction and oxygen inhibition. Cationic photoinitiators have been studied extensively[1)], and utilized to develop several new and commercially important technologies. Anionic photoinitiators have also been developed in similar ways as cationic photoinitiators. However, despite numerous studies of cationic photoinitiators, only a few have been reported on the photo-induced base formation[2)].

The first example of a photobase generating compound was reported by Willson et al.[3)] in 1987. The photochemical formation of free amines from chemical systems such as cobalt-amine complex[3)], benzoyl formic acid amine salt[4)], urethane[5)], o-nitrobenzoyl carbamate[6)], oxime-ester[7)], blocked aniline[8)], and benzoin carbamate[9)] derivatives has been reported. The photobase generators have been applied to photoresists[10-11)], cross-linking reagents for epoxy resins[7, 12)], imidation catalysts[13-14)], and surface modification[15)].

 CCC 1022-1360/99/$ 17.50+.50/0

Scheme 1:

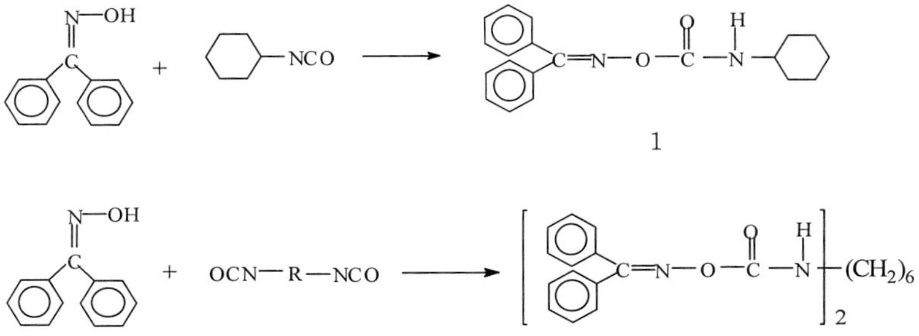

Oxime-urethane groups have been used as precursors to isocyanate groups[16]. We found that the photolysis of oxime-urethane derivatives led to the formation of amines which induce cross-linking reaction of epoxy resins upon heating. In this paper we describe the photochemical base formation from oxime-urethane derivatives and its application to some polymeric systems.

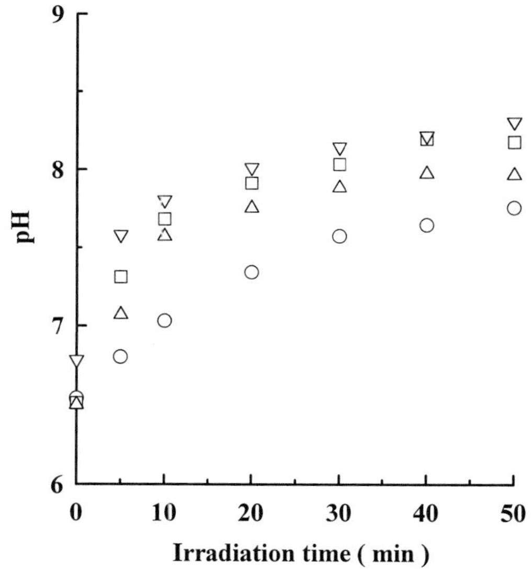

Fig. 1. pH changes of oxime-urethane derivatives as a function of irradiation time in acetonitrile-water solution: (O) compound **1**; (Δ) compound **1** with benzophenone; (□) compound **2**; (∇) compound **2** with benzophenone.

Photogeneration of amine from oxime-urethane derivatives[17)]

Oxime-urethane derivatives were prepared from the reaction of cyclohexyl isocyanate or hexamethylenediisocyanate with benzophenone oxime, as shown in Scheme 1.

Photoinduced amine formation from oxime-urethane derivatives was confirmed by GC, HPLC, and pH changes. Fig. 1 shows pH changes of oxime-urethane derivatives in acetonitrile-water solution. The pH changed from 6.5 to 8.2 upon irradiation with 254 nm UV light. The addition of benzophenone augmented the pH changes, which implies that this photochemical reaction may take place in its excited triplet state. The photobase generation efficiency for bifunctional oxime-urethane derivative **2** was greater than that of mono-functional derivative **1**.

Scheme 2 shows a possible reaction mechanism for the photogeneration of amine from oxime-urethane derivatives. It is expected that photolysis of oxime-urethane derivatives would lead to the homolytic cleavage of N-O bond, resulting in an iminyl radical and carbonate radical pair. The iminyl radical can undergo hydrogen abstraction to form an imine, or dimerization to form an azine. The carbonate radical can undergo decarboxylation followed by hydrogen abstraction to form an amine.

Scheme 2:

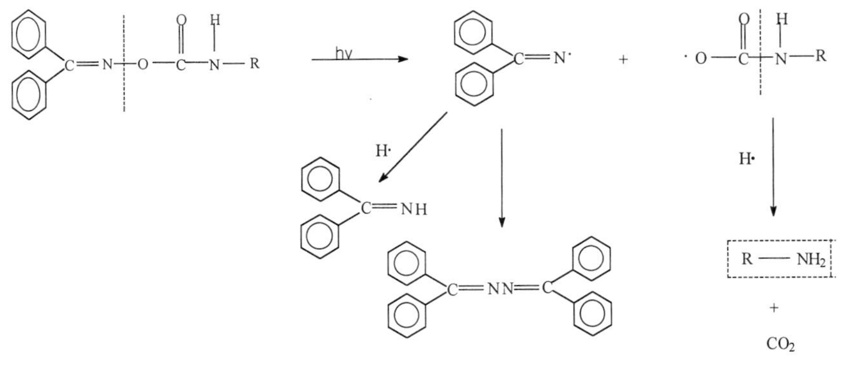

Laser flash photolysis study[18)]

The photochemical and photophysical phenomena of compound **2** was investigated with laser flash photolysis to understand its photodecomposition mechanism in anionic polymerization reactions. Scheme 3 shows the summary of our laser flash photolysis study of compound **2**. Irradiation of 266 nm laser pulse to compound **2** results in the transition from ground state

Scheme 3:

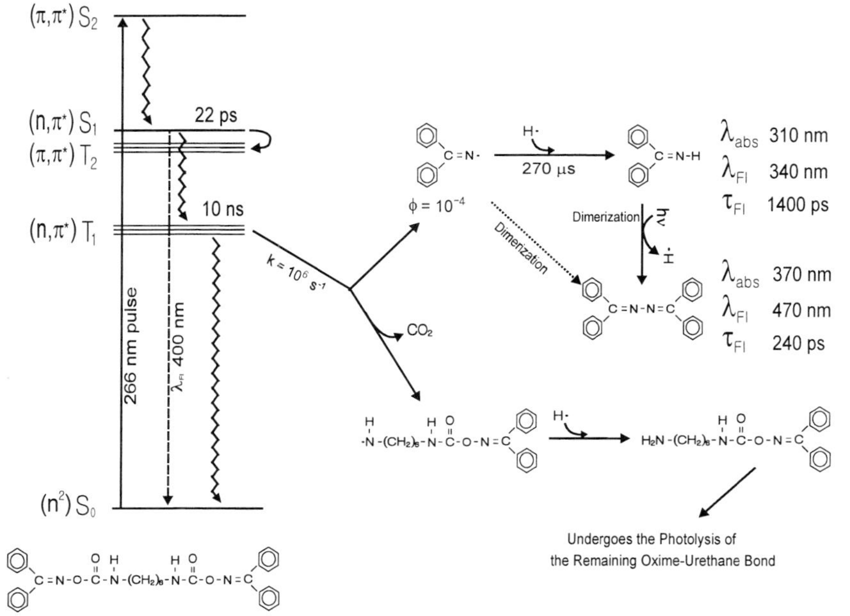

to the (π, π*) S_2 state, followed by a fast internal conversion to the (n, π*) S_1 state. In 22 ps the population of the S_1 state undergoes a rapid intersystem crossing to the (π, π*) T_2 state that happens to lie near the S_1 state in energy. The (n, π*) T_1 state, populates immediately via internal conversion from T_2 state, is then depopulates in the time scale of 10 ns via intersystem crossing to the S_0 ground state.

About 2 % of the T_1 state population tend to undergo decarboxylation within its decay time, generating dehydrogenated iminyl radical and aminyl radical. Abstracting a hydrogen atom within the time frame of 270 μs, the iminyl radical converts mainly into diphenyl-1-imine which, upon irradiation, dehydrogenates and dimerizes to yield benzophenone azine. Meanwhile, the aminyl radicals abstract H to generate benzophenoneoxime-urethanyl hexamethyleneamine, which raises solution pH or initiates thermal curing reaction of epoxy resin as a photobse.

Cross-linking reaction of epoxy resin photoinduced by oxime-urethane derivatives[17)]

Amines are very important reagents for thermal cross-linking reaction of epoxy resins in adhesives, paints, and coatings industry. However, a mixture of epoxy resins with aliphatic amines is very unstable due to the high reactivity of aliphatic amines. Therefore, light induced amine formation from blocked amine groups is a very useful technique for the formulations of epoxy resins which can lead to a new industrial technology.

The thermal cross-linking reaction of poly(glycidyl methacrylate) (PGMA) films containing compound **1** as a function of heating temperature showed that the PGMA films without the addition of compound **1** were not cross-linked upto 120 ^{0}C, for heating time of 10 min. However, the insoluble fraction increased with heating temperature and irradiation time in the presence of compound **1**. Slight thermal cross-linking behavior was observed for non-irradiated PGMA films containing compound **1**. This may be due to the oxime-urethane groups which produce isocyanate groups upon heating, that can react with epoxy groups in PGMA to form cross-links.

Fig. 2 shows the effect of heating time on the thermal cross-linking reaction of irradiated PGMA films containing 5 wt.-% of compound **1** at 100 ^{0}C. The insoluble fraction of non-irradiated PGMA films increased slightly with heating time. However, the insoluble fraction

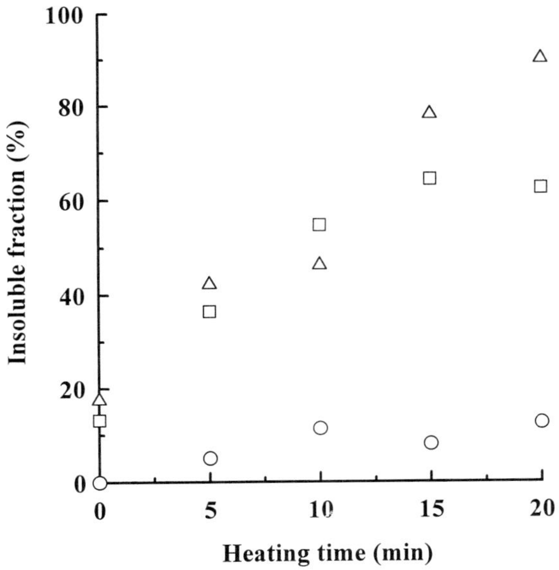

Fig. 2. Effects of heating time on thermal cross-linking reaction of irradiated PGMA films containing 5 wt.-% of compound 1: (o) no irradiation; (□) irradiated for 5 min; (Δ) irradiated for 10 min. Heating temperature, 100 °C.

Scheme 4:

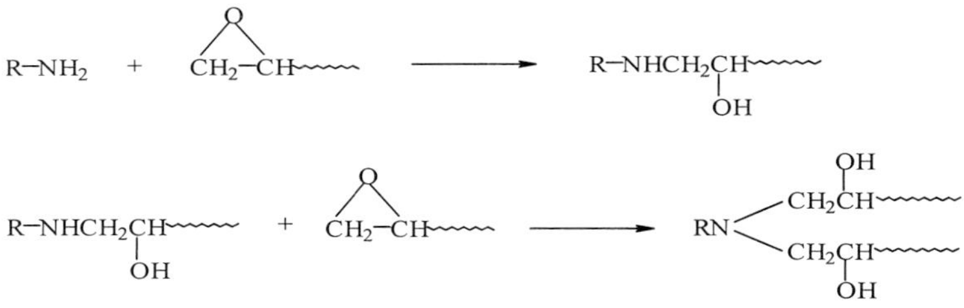

of PGMA films after irradiation increased with heating and irradiation time. After 20 min of heating, an appreciable increase in the insoluble fraction, upto 90%, was observed for the PGMA films irradiated for 10 min. This indicates that thermal cross-linking of PGMA films was induced by the amines which were produced by the photolysed oxime-urethane derivatives.

The insoluble fraction of PGMA films containing two different oxime-urethane derivatives as a function of irradiation time showed that bifunctional derivative **2** was more efficient than monofunctional derivative **1** in inducing thermal cross-linking of PGMA films.

Amine-induced cross-linking mechanism of PGMA is shown in Scheme 4. Photochemically produced amines from oxime-urethane derivatives attack epoxy groups of PGMA to form secondary amines, which in turn, react with other epoxy groups to form cross-links.

Scheme 5:

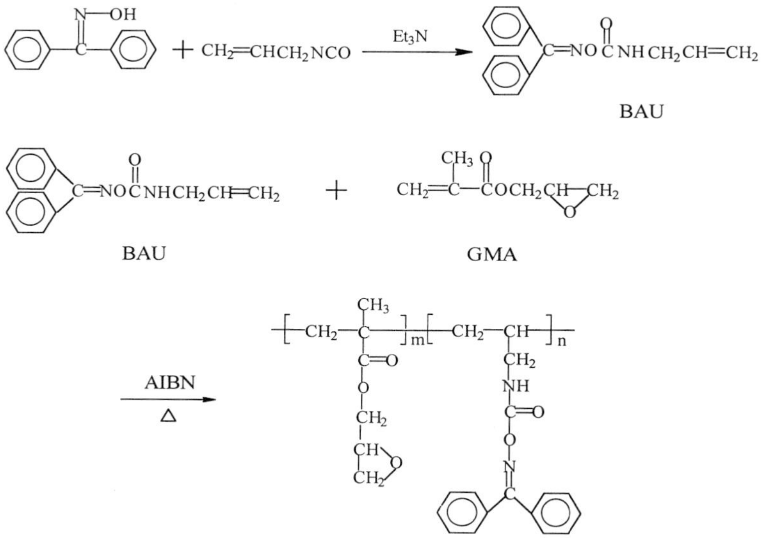

Photoinduced cross-linking reaction of epoxy resin containing oxime-urethane groups[19)]

The idea of the photobase generation from oxime-urethane derivatives was extended to the thermal cross-linking reaction of epoxy resins containing oxime-urethane groups. The synthetic route for the preparation of glycidyl methacrylate copolymers containing oxime-urethane groups is shown in Scheme 5. Benzophenoneoxime allyl urethane (BAU), a monomer containing oxime-urethane group, was prepared from benzophenoneoxime and allyl isocyanate. Copolymers containing oxime-urethane groups was prepared by the copolymerization of BAU and GMA.

Table 1 shows the physical properties of the copolymers which were prepared by varying the molar ratio of BAU from 9 to 33%. The content of BAU units in the copolymer was in the range of 0.7 – 3.0 %. Intrinsic viscosity of the copolymers was in the range of 0.23 – 0.32. Molecular weight and intrinsic viscosity decreased with the amount of BAU units in the copolymer. Tg of these copolymers ranged between 85 –90 °C, with Tm of approximately 280 °C.

Table 1. Physical properties of the copolymers

Copolymer[a]	Mole % of BAU		Yield	Intrinsic	Tg [c]	Molecular weight		
	in feed	in copolymer[b]	(%)	viscosity	(°C)	Mn	Mw	Dispersity
I	9.1	0.7	55	0.32	87	138,000	166,000	1.20
II	16.7	1.2	60	0.26	90	129,000	162,000	1.26
III	33.3	3.1	59	0.23	85	124,000	162,000	1.31

[a] Polymerization was carried out with 1 wt% of AIBN in 3 ml of THF at 60 °C for 24 hours.
[b] The content of BAU units in the copolymer was determined from 300 MHz NMR spectra.
[c] Obtained from DSC thermogram under N_2 atmosphere with scan speed of 10 °C/min.

The photo-crosslinking reaction of the copolymers containing epoxy and oxime-urethane groups was studied by measuring the insoluble fraction of the copolymer films under various reaction conditions.

Fig. 3 shows the effect of heating temperature on the insoluble fraction of PGMA and copolymer I films in THF before and after irradiation. The PGMA films thermally cross-linked up to 37 % at 140 °C. The insoluble fraction of non-irradiated copolymer I film containing 0.7 % of BAU units increased about twice that of PGMA. The insoluble fraction of the copolymer film after irradiation was 25-30 % higher than that of non-irradiated

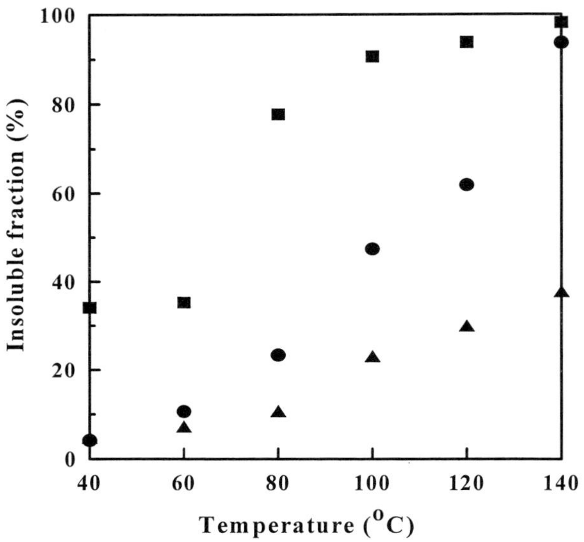

Fig. 3. Effect of heating temperature on the insoluble fraction of PGMA (▲) and copolymer I film in THF before (●) and after (■) irradiation for 40 min with heating time for 40 min.

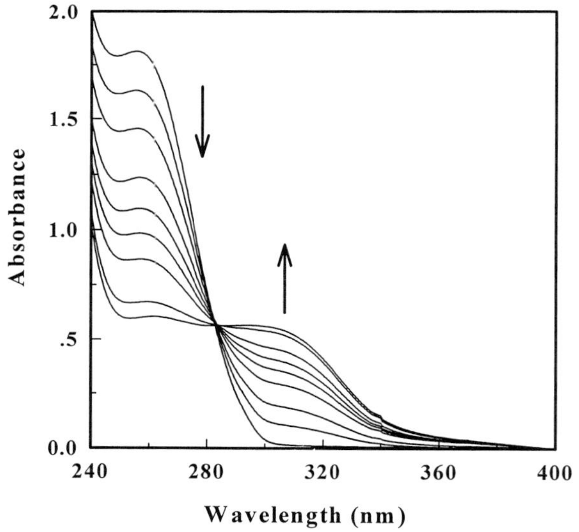

Fig. 4. UV absorption spectral changes of copolymer I in THF with irradiation time.

copolymer film at 40 – 60 °C. However, the difference in the insoluble fraction of irradiated and non- irradiated copolymer films increased up to 55 % at 80 °C. Thus, the higher insoluble fraction observed for the irradiated copolymer films can be attributed to the thermal cross-linking reaction of copolymers catalyzed by the amines produced from photolysed oxime-urethane groups.

The insoluble fraction of copolymer films in THF as a function of irradiation time at different heating temperatures was studied. The insoluble fraction increased with irradiation time, heating temperature and the content of BAU units in the copolymer. The maximum insoluble fraction of copolymer I film was 35 %, below 60 °C, but it increased up to 95 % above 80 °C at the irradiation time of 40 min. Thus there was a large difference in the insoluble fraction of copolymer films between 60 – 80 °C. This can be explained by the mobility difference of photogenerated amino groups in the polymer chain at the temperatures below and above Tg.

UV absorption spectral changes with irradiation was observed in order to study the photochemical reaction of copolymers. Fig. 4 shows UV absorption spectral changes of

Scheme 6:

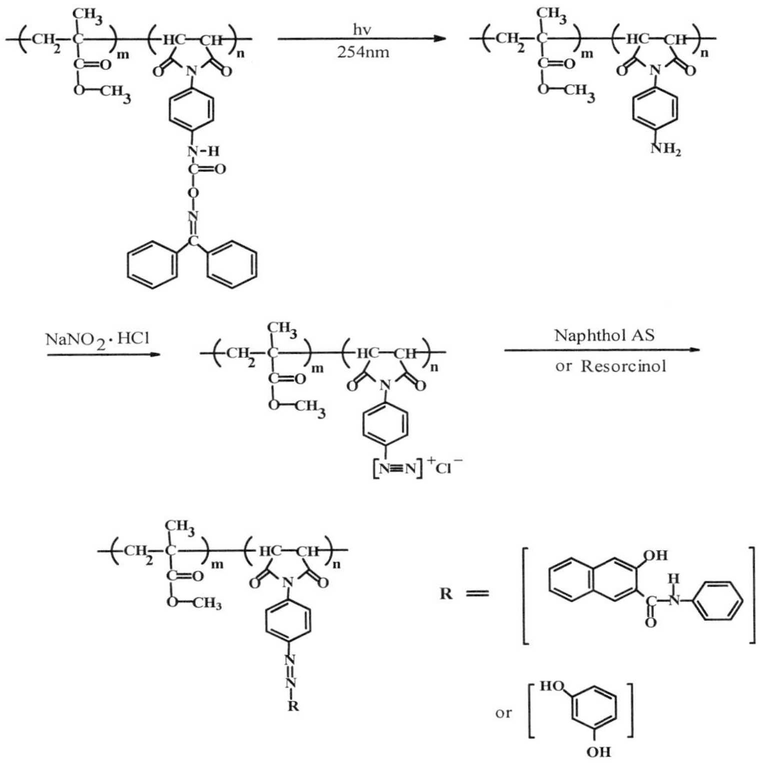

copolymer I film upon irradiation with UV light of 254 nm for various irradiation time. An isosbestic point appeared at 283 nm. The absorption band at wavelengths shorter than the isosbestic point decreased, while that at wavelengths longer than the isosbestic point increased. The decrease in absorbance at wavelengths shorter than 283 nm is due to the photodecomposition of oxime-urethane groups while the increase at wavelengths longer than 283 nm is due to the formation of azine type compounds.

UV and IR absorption spectral studies indicate that oxime-urethane groups in the polymer chain photodecomposed to form pendant amines, which in turn effectively catalysed the thermal cross-linking reaction of epoxy groups in the copolymer chain.

Applications in image recording material

Scheme 6 shows a chemical reaction which displays a possible application of the photobase generating polymer as an image recording material. The irradiation of the MMA copolymer containing oxime-urethane groups leads to the formation of aromatic pendant amines in the irradiated parts. Further reactions with sodium nitrite generate diazonium salts, which are treated with phenol derivatives such as resorcinol or naphthol AS, resulting in the formation of diazo groups. Various colors were developed depending on the phenolic compounds.

Fig. 5 shows an example of image recording using this polymer. The images in this figure represent a raven with three legs, symbolizing 'the sun' in Koguruh dynasty about 1000 years ago in Korea. The color red was developed from treatment with naphthol AS while yellow was developed from treatment with resorcinol.

Fig. 5. An example of image recording. The color red was developed in the irradiated parts after diazotization followed by treatment with naphthol AS on the polymer surface.

Conclusion

The photolysis of oxime-urethane derivatives leads to the formation of amine which induces cross-linking reaction of epoxy resin upon heating. Epoxy copolymers containing oxime-urethane groups effectively cross-linked, induced by the photogenerated pendant amine. Thus, polymers which can generate photochemical aromatic amines can be utilized as an image recording material.

Acknowledgement

This work was financially supported by the Korea Science and Engineering Foundation (Grant No. 98-05-01-03-01-3). DJJ acknowledges the Center of Molecular Science, Korea and the Basic Science Research Institute Program.

References

1) See e.g., J. V. Crivello, "*Radiation Curing in Polymer Science and Technology*", Vol. II: Photoinitiating System, eds by J. P. Fouassier, J. F. Rabek, Elsevier Applied Science, London and New York 1993, p. 435.
2) K. H. Chae, *Polym. Sci. & Tech.*, **8**, 421 (1997).
3) C. Kutal, C. G. Willson, *J. Electrochem. Soc.*, **134**, 2280 (1987).
4) T. Fuji, *Jpn. KoKai Tokyo Koho*, 63189427: *Chem. Abstr.*, **110**, 116014 (1988).
5) J. F. Cameron, J. M. J. Frechet, *J. Org. Chem.*, **55**, 5919 (1990).
6) J. F. Cameron, J. M. J. Frechet, *J. Amer. Chem. Soc.*, **113**, 4303 (1991).
7) K. Ito, Y. Shigeru, Y. Kawata, K. Ito, M. Tsunooka, *Can. J. Chem.*, **73**, 1924 (1995).
8) T. Nishikubo, E. Takehara, A. Kameyama, *Polym. J.* **25**, 421 (1993).
9) J. F, Cameron, C. G. Willson, J. M. J. Frechet, *Polym. Mat. Eng. Sci.*, **74**, 323 (1996).
10) M. R. Winkle, K. A. Graziano, *J. Photopol. Sci. & Tech.*, **3**, 419 (1990).
11) M. J. Frechet, S. Matuszczak, S. M. Lee, J. Fahey, *Polym. Eng. Sci.*, **32**, 1471 (1992).
12) T. Nishikubo, E. Takehara, A. Kaneyama, *Polym. J.* **25**, 421 (1993).
13) D. R. McKean, G. M. Wallraff, W. Volksen, N. P. Hacker, M. I. Sanchez, J. W. Labadie, *Polym. Mater. Sci. Eng.*, **66**, 237 (1992).
14) J. M. J. Frechet, J. F, Cameron, C, M. Chung, S. A. Haque, C. G. Willson, *Polym. Bull.*, **30**, 369 (1993).
15) K. H. Song, A. Iwamoto, M. Tsunooka, M. Tanaka, *J. Appl. Polym. Sci.*, **39**, 1769 (1990).
16) T. Anagnostou, E. Jaul, *J. Coatings Technol.*, **53**, 35 (1981).
17) K. H. Chae, *Macromol. Rapid Commun.*, **19**, 1 (1998).
18) H. Hwang, D.-J. Jang, K. H. Chae, *J. Phys. Chem.* submitted, 1998.
19) K. H. Chae, H. B. Song, *Polym. Bull.*, **40**, 667 (1998).

Highly Efficient Template-based Preparation of Shape-persistent Macrocyclics

S. Höger

Max-Planck-Institute for Polymer Research, Ackermannweg 10, 55128 Mainz, Germany

SUMMARY: The synthesis of shape-persistent macrocycles via oxidative dimerization or trimerization of rigid bisacetylenes (via Glaser coupling) is described. To enhance the yield of the desired target structures, the precursors are temporarily covalently bound to a template resulting in an intramolecular coupling instead of an intermolecular coupling. As a result, product yields up to over 90 % are observed.

Introduction

Macrocyclic structures with a rigid, non-collapsible backbone and interior sizes of several nanometers are interesting objects in the field of Supramolecular Chemistry[1]. Besides interesting aggregation phenomena in solution[2] and in the solid state[3], especially their ability to act as host molecules for relatively large guest molecules has made them desirable target strtuctures[4]. However, only a limited number of reports on these molecules is available, probably because of the huge effort often required in their preparation.

Two extreme strategies for their synthesis can be distinguished. In the first startegy, the "shotgun synthesis", formation of the oligomeric precursor and its cyclization is performed in a one-pot procedure, leading to a wide variety of cyclic and non-cyclic oligomers and polymers (scheme 1a)[5].

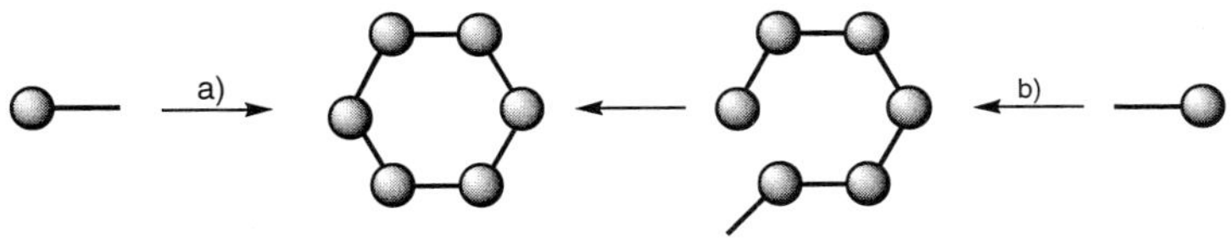

Scheme 1: Different approaches to shape-persistent macrocycles

In contrast, if the precursor of the cyclization reaction is prepared repetitively in a seperate vessel prior to its cyclization, the intramolecular cyclization under (pseudo) high-dilution conditions generally gives high yields of the ring (scheme 1b). However, the synthesis of the precursor is often a time consuming multi-step procedure. Several examples exist in the

 CCC 1022-1360/99/$ 17.50+.50/0

literature which allow comparisons between the effort needed to prepare the precursor and the cyclization yield[6]. From these data it seems that macrocyclizations find themselves in a paradox: *either easily available starting materials are used for cyclizations resulting in a low product yield, or high yields and simplified purifications are obtained at the cost of a time-consuming precursor synthesis.*

Results and Discussion

In our previous investigations of the synthesis and properties of shape-persistent macrocyclic amphiphiles with a switchable arrangement of the polar and non-polar functionalities, we used as cyclization step the intermolecular oxidative Glaser coupling of the rigid bisacetylenes **1** ("half-rings") under pseudo high-dilution conditions (scheme 2a)[7].

a)

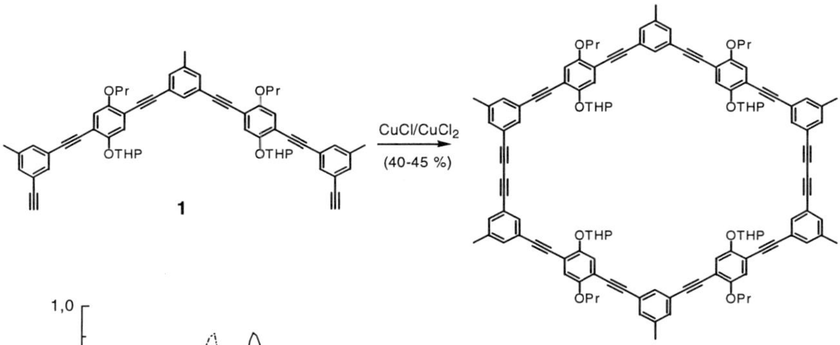

b)

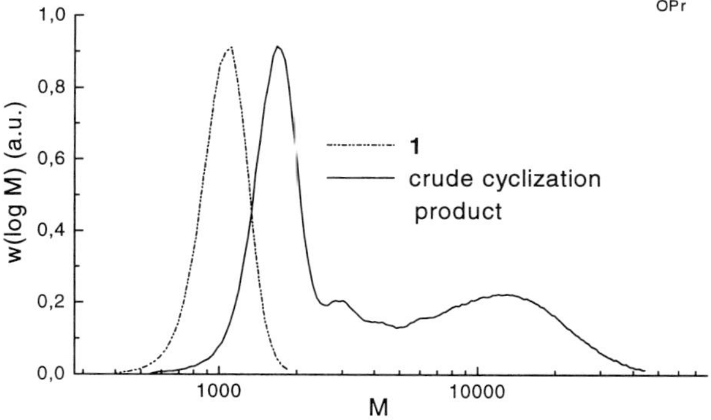

Scheme 2: Cyclization of rigid "half-rings" under pseudo high-dilution conditions

Gel permeation chromatography (GPC) has been proven to be a valuable tool for the investigation of the reaction products (the molecular weight is obtained from polystyrene calibration of the size exclusion chromatography (SEC) columns). Scheme 2b shows the elution diagrams of **1** and of the crude cyclization product which is usually obtained in more than 95 % yield. According to this data the cyclic dimer **2** is formed in 60 - 65 % yield along with cyclic trimer and cyclic or non-cyclic oligomers and polymers. **2** is obtained by simple recrystallization from dichloromethane in 40 - 45 % yield.

One approach to simplify precursor synthesis is the preparation of smaller building blocks such as **3** ("thirds of a ring") (scheme 3) in which prior to its oxidative cyclization the polar alcohol functionality must be protected, here as the benzoate **4**. Intermolecular cyclization under the same conditions as used for the cyclization of **1** forms a crude product (> 95 % yield) containing mostly polymeric material. The cyclic trimer is formed in only 20 - 25 % yield (according to GPC analysis) [8].

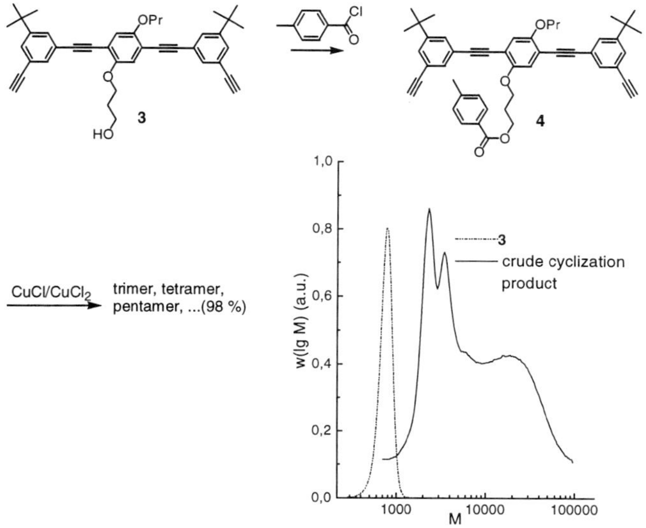

Scheme 3: Cyclization of rigid "thirds of a ring " under pseudo high-dilution conditions

This relatively low yield is not surprising, since the joining of three building blocks in a cyclic compound is less probable than the joining of only two building blocks. However, if a different protective group for the alcohol functionality is chosen and the trimesinic acid ester **5** is prepared (under Mitsunobu conditions [9]) (scheme 10a), the cyclization of this template-bound hexaacetylene **5** under identical conditions yields a crude product which contains more than 95 % of the template-bound cyclic trimer **6** which is isolated in 94 % yield. Base catalysed cleavage of the ester bonds gives the desired macrocycle **7** in nearly quantitative yield [8].

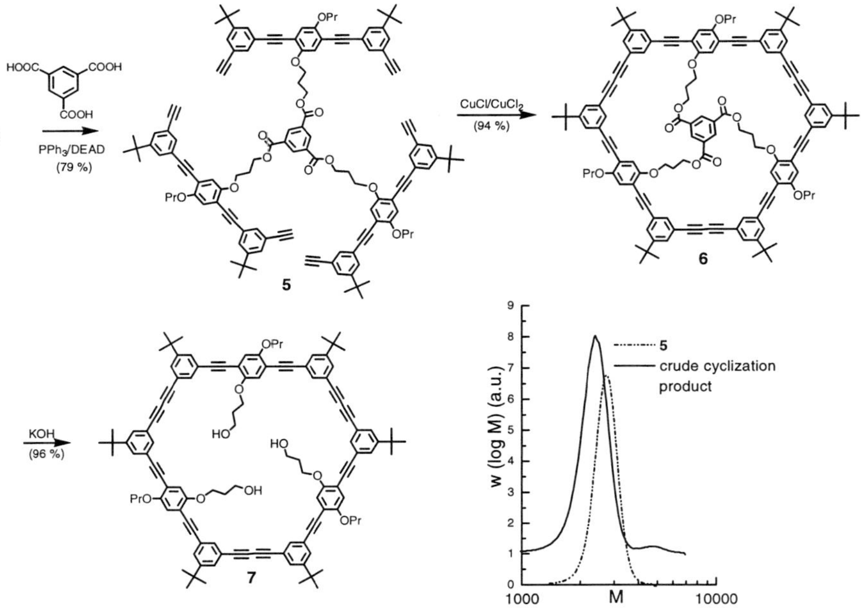

Scheme 4: Template-directed synthesis of shape-persistent macrocycles

This was the first time that we were able to obtain a shape-persistent macrocycle with a switchable arrangement of the amphiphilic portions of the ring in high yield using relatively small and easily available starting materials. Moreover, it has been demonstrated by the successful cyclization of homologues of **5** with a dramatically enlarged distance between the template and the rigid acetylenic parts of the molecule that for a high product yield it is not necessary to arrange the bisacetylenes in a particular preorganised geometry. Rather, the high cyclization yield can be attributed to an overall low reactant concentration as a result of the high-dilution conditions together with a high local concentration of terminal acetylenes as a result of the intramolecular nature of the coupling reaction [10,11]. With the template strategy it is also possible to prepare macrocycles with polar groups pointing to the outside, valuable A_2-monomers[12].

Changing the succession of the steps in the synthesis of the templated precursors gives us even more opportunities to modify the functional groups within the ring (scheme 5)[13]. Instead of preparing the bisacetylenes first and attaching them to the template in a separate step, the complete synthesis of the templated tetraacetylenes is performed *at the template*. The corner pieces of the ring containing the functional groups (**8**) (which now points to the inside of the final structure leading to compounds with intraannular functional groups[14]) are first attached to the template (**9**) to give the templated tetraiodide **10**, the starting material for **15**, a nanometer-size shape-persistent structure with a convergent arrangement of functional groups.

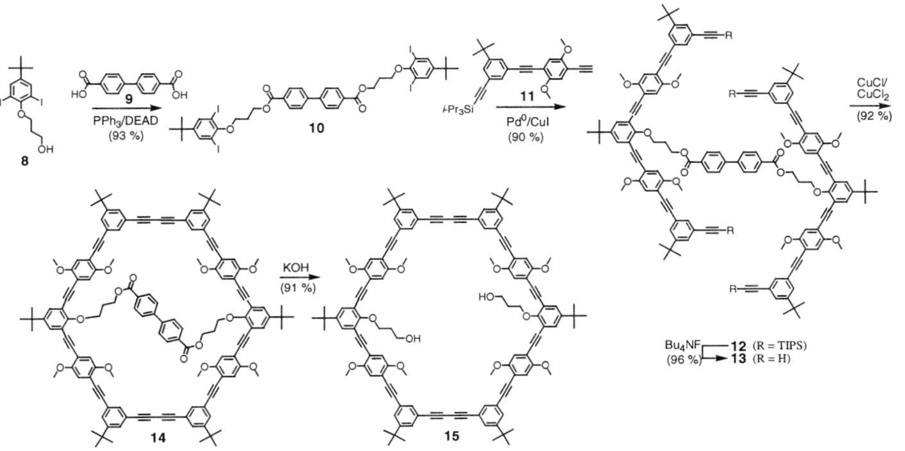

Scheme 5: Template-directed synthesis of macrocycle containing intraannular substituents

There are several advantages to this strategy. First only a few steps are necessary to prepare the templated bisacetylenes. Second, all these steps give relatively high yields because the molecular weights and the physical properties of the reactants and products change so dramatically in every step that product purification is simplified. And last, but of perhaps highest significance, is the fact that macrocycles with two different functional groups can be prepared by the same methodology using the non-symmetrical tetraiodide **16** (scheme 6). The major advantage of this procedure is that all structural information necessary to obtain the non-symmetrical macrocycle **17** is introduced on the level of the low molecular weight precursors which are much easier to handle than high molecular weight ring precursors.

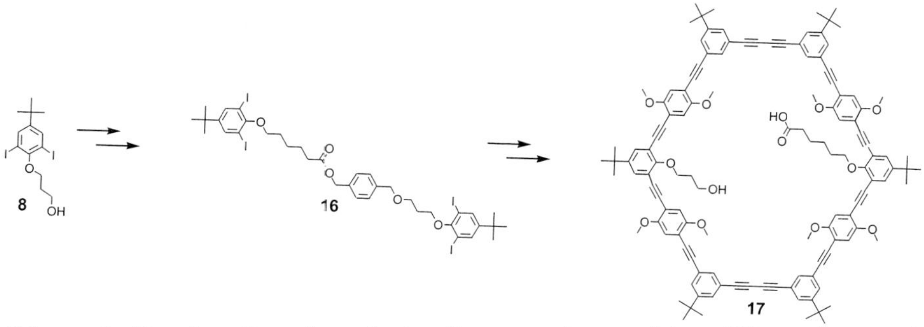

Scheme 6: Template-directed synthesis of macrocycle containing different intraannular substituents

References and notes

1) J.L. Atwood, J.E.D. Davies, D.D. Macnicol, F. Vögtle, (ed.), *Comprehensive Supramolecular Chemistry, vol. 3, Cyclodextrines*, Elsevier Science Ltd., Oxford, **1996**;

2) (a) J. Zhang, J.S. Moore, *J. Am. Chem. Soc.* **1992**, *114*, 9701; (b) A.S. Shetty, J. Zhang, J.S. Moore, *J. Am. Chem. Soc.* **1996**, *118*, 1019; (c) Y. Tobe, N. Utsumi, K. Kawabata, K. Naemura, *Tetrahedron Lett.* **1996**, *37*, 9325; (d) Y. Tobe, N. Utsumi, A. Nagano, K. Naemura, *Angew. Chem., Int. Ed. Engl.* **1998**, *37*, 1285.

3) (a) J. Zhang, J.S. Moore, *J. Am. Chem. Soc.* **1994**, *116*, 2655; (b) S. Höger, C. Tschierske, K. Bonrad, V. Enkelmann, in preparation.

4) For macrocyclic host molecules based on the phenyl-ethynyl backbone see e.g. (a) H.L. Anderson, J.K.M. Sanders, *J. Chem. Soc. Chem. Commun.* **1989**, 1714; (b) L.G. Mackeay, H.L. Anderson, J.K.M. Sanders, *J. Chem. Soc. Chem. Commun.* **1992**, 43; (c) H.L. Anderson, J.K.M Sanders, *J. Chem. Soc. Chem. Commun.* **1992**, 946; (d) S.Anderson, U. Neidlein, V. Gramlich, F. Diederich, *Angew. Chem., Int. Ed. Engl.* **1995**, *34*, 1596; (e)D.L. Morrison, S. Höger, *J. Chem. Soc. Chem. Commun.* **1996**, 2313.

5) (a) L.T. Scott, M.J. Cooney, D. Johnels, *J. Am. Chem. Soc.* **1990**, *112*, 4054; (b) L.T. Scott, M.J. Cooney, in *Modern Acetylene Chemistry* (ed. P.J. Stang, F. Diederich), VCH, Weinheim, **1995**, pp. 321.

6) For example, see for a comparison: (a) H. A. Staab, K. Neunhoeffer, *Synthesis* **1974**, 424 and (b) J. S. Moore, J. Zhang, *Angew. Chem. Int. Ed. Engl.* **1992**, *31*, 292, as well as (c) H.

Staab, F. Binning, *Chem. Ber.* **1967**, *100*, 293; and (d) V. Hensel, K. Lützow, J. Jakob, K. Gessler, W. Saenger, A.-D. Schlüter, *Angew. Chem. Int. Ed. Engl.* **1997**, *36*, 2654.

7) (a) S. Höger, V. Enkelmann, *Angew. Chem. Int. Ed. Engl.* **1995**, *34,* 2713; (b) S. Höger, A.-D. Meckenstock, S. Müller, *Chem. Eur. J.* **1998**, *4*, 2421.

8) S. Höger, A.-D. Meckenstock, H. Pellen, *J. Org. Chem.* **1997**, *62*, 4556.

9) (a) O. Mitsunobu, *Synthesis* **1981**, 1; (b) D.L. Hughes, *Org. React.* **1992**, *42*, 335.

10) For recent articles dealing with template reactions see e.g.: (a) B. Dietrich, P. Viout, J.-M. Lehn, *Macrocyclic Chemistry*, VCH, Weinheim, **1993**. (b) S. Anderson, H.L. Anderson, J.K.M. Sanders, *Acc. Chem. Res.* **1993**, *26*, 469; (c) R. Hoss, F. Vögtle, *Angew. Chem., Int. Ed. Engl.* **1994**, *33*, 375; (d) J.L. Atwood, J.E.D. Davies, D.D. Macnicol, F. Vögtle, (ed.), *Comprehensive Supramolecular Chemistry, vol.* 9, Elsevier Science Ltd., Oxford, **1996**.

11) For some reviews and examples on the use of covalently attached templates in organic synthesis see e.g.: (a) R. Breslow, *Acc. Chem. Res.* **1980**, *13*, 170; (b) K.S. Feldman, Y.B. Lee, *J. Am. Chem. Soc.,* **1987**, *109*, 5850; (c) J.W. Gillard, R. Fortin, E.L. Grimm, M. Maillard, M. Tjepkema, M.A. Bernstein, R. Glaser, *Tetrahedron Lett.*, **1991**, *32*, 1145; (d) J.-I. Hong, S.K. Namgoong, A. Bernardi, W.C. Still, *J. Am. Chem. Soc.,* **1991**, *113*, 5111; (e) S. Valverde, A.M. Gómez, J.C. López, B. Herradón, *Tetrahedron Lett.* **1996**, *37*, 1105; (f) F. Cardullo, L. Isaacs, F. Diederich, J.-P. Gisselbrecht, C. Boudon, M. Gross, *J. Chem. Soc. Chem. Commun.* **1996**, 797; (g) H. Weizman, J. Libman, A. Shanzer, *J. Am. Chem. Soc.,* **1998**, *120*, 2188.

12) S. Höger, A.-D. Meckenstock, *Tetrahedron Lett.* **1998**, *39*, 1735.

13) S. Höger, A.-D. Meckenstock, *Chem. Eur. J.* **1999**, in press.

14) For early reports of the concept of intraannular functional groups see e.g.: (a) E. Weber, F. Vögtle, *Angew. Chem. Int. Ed. Engl.* **1974**, *13,* 149; (b) E. Weber, F. Vögtle, *Chem. Ber.* **1976**, *109*, 1803.

Acknowledgement: Financial support for a part of this work by the Fonds der Chemischen Industrie is gratefully acknowledged. The author thanks the German Chemical Society for Support by the Hermann Schnell-Stiftung.

Macromol. Symp. ***142***, *193–204 (1999)*

Soluble Polyimides Containing Alicyclic Structures

Kil-Yeong Choi*, Mi Hie Yi

Advanced Materials Division, Korea Research Institute of Chemical Technology, P.O.Box 107, Yusong, Taejon, 305-600, Korea

SUMMARY : A series of novel aromatic diamines containing kinked cycloalkane structures between two phenyl rings were synthesized by HCl-catalyzed condensation reaction of excess aniline and corresponding cycloalkanone derivatives. The structures of the diamines were indentified by ^1H NMR, ^{13}C NMR, FT-IR spectroscopy and elemental analysis. The polyimides were synthesized from the obtained diamines with various aromatic dianhydrides by one-step polymerization in *m*-cresol. The polymerization was conducted for 6~8 h with refluxing, which was enough to obtain the polymers with high molecular weight. The inherent viscosities of the resulting polyimides were in the range of 0.37~1.66 dl/g. All polymers were readily soluble in common organic solvents such as chloroform, tetrachloroethane, dimethylacetamide, etc. and the glass transition temperatures were observed at 290 to 372°C. UV-visible spectra were obtained to measure the transparency of polymer films. Most of the polymers showed high transmission above 90 % in the wavelength of 450 ~ 600 nm.

Introduction

Polyimides have been noted for their excellent characteristics such as thermo-mechanical property, electrical properties, and chemical resistance[1]. However, wholly aromatic polyimides with rigid structure were usually insoluble and intractable, which resulted in difficulties of processability. Furthermore, it was well known that intermolecular charge-transfer complexes(CTCs) formed between the electron-rich and electron-deficient moieties of polymer chains would be caused for dark color and high dielectric constant. Therefore, significant synthetic efforts have been focused on the synthesis of soluble, colorless and processable polyimides without the sacrifice of their excellent properties. One approach was to introduce the flexible linkages like amides and ether groups, or the incorporation of bulky pendant groups[2]. Another approaches were the incorporation of alicyclic structures into the polymer backbone reported by Kurosaki et al.[3-10] In this study, we would like to introduce the preparation of a class of soluble polyimides with high molecular weight from several kinds of aromatic diamines containing alicyclic moieties by one step polymerization in *m*-cresol. We have prepared a series of novel aromatic diamines containing various alicyclic structures

 CCC 1022-1360/99/$ 17.50+.50/0

between two phenyl rings and synthesized the new soluble polyimides from the prepared diamines and various aromatic dianhydrides. Also, the general properties of these polymers have been studied and compared with those of conventional aromatic ones [11-12].

Experimental

Materials. *m*-Cresol was freshly distilled under reduced pressure. *N*-methyl-2-pyrrolidone (NMP) was dried over calcium hydride for 12 h and distilled under reduced pressure. Pyromellitic dianhydride (PMDA, 99.2 %, Daicel Chemical Industry, Ltd.), 3,3'4,4'-benzophenonetetracarboxylic dianhydride (BTDA, 98.6 %, Daicel Chemical Industry, Ltd.), and 3,3'4,4'-biphenyltetracarboxylic dianhydride (BPDA, 99.6%, Mitsubishi Kasei) were used without further purification, because they were of highly purified grade for polyimide synthesis. 2,2-Bis(3,4-dicarboxyphenyl)hexafluoropropane dianhydride (6FDA, Tokyo Chemical Industry Co., Ltd., >98%) and 3,3',4,4'-tetracarboxydiphenyl oxide dianhydride(ODPA, Tokyo Chemical Industry Co., Ltd., >98%) were used as received. Aniline (Aldrich Chemical Co., Inc., 99%, bp ;184 °C), cyclohexanone (Tokyo Chemical Industry Co., Ltd., >98%, bp;157°C), 4-methylcyclohexanone (Tokyo Chemical Industry Co., Ltd., >97%, bp;171°C), 3,3,5-trimethylcyclohexanone(Tokyo Chemical Industry Co., Ltd., >98%, bp;190°C) , 4-ethylcyclohexanone (Tokyo Chemical Industry Co., Ltd., >98%), 4-t-butylcyclohexanone (Tokyo Chemical Industry Co., Ltd., >98%, mp;50°C) and 4-n-pentylcyclohexanone(Tokyo Chemical Industry Co., Ltd., >90%) were used as received.

Measurements. Infrared spectra were obtained with a Bio-Rad Digilab Division FTS-165 FT-IR spectrophotometer and All NMR spectra were taken on a Bruker AMX 500 using $CDCl_3$ or dimethyl sulfoxide-d_6 (DMSO-d_6) as solvents. UV-Vis spectra were recorded on a Shimadzu UV 2100 UV-visible spectrotometer. Mass spectra and elemental analysis were measured with Jeol JMS-DX 303 mass spectrometer and FISONS Instruments EA-1108, respectively. The inherent viscosities of polymers were measured with a Ubbelohde viscometer at a concentration of 0.5 g/dL in *m*-cresol or NMP at 30 °C. The molecular weight of polymers were measured by Waters Model 150C ALC/GPC(Gel Permeation Chromatography) equipped with μ-Styragel columns in chloroform solution at 40 °C. The melting points of the prepared monomers were determined by DSC. Thermogravimetric analysis (TGA) was done with a Perkin Elmer TGA 7 at heating rate of 20 °C/min under nitrogen atmosphere. Differential scanning calorimeter (DSC) analysis was performed on a Du Pont model 910.

Monomer synthesis[11-13)]

1,1-Bis(4-aminophenyl)cyclohexane(BACH). BACH was prepared by the condensation reaction of cyclohexanone and excess aniline as follows.[14-15)] To a solution of cyclohexanone (40.0 g, 0.41 mol) in 145 mL of 35 % HCl aqueous solution in a 1 L autoclave equipped with a mechanical stirrer was added excess aniline (156.3 g, 1.68 mol) and the mixture was stirred at 120 °C for 20 h. After cooling , the solution was made basic with NaOH aqueous solution to pH 10 and the oily layer was separated and distilled to remove the unreacted excess aniline. The residual crude product was recrystallized from benzene to afford 86.4 g (79.4 % yield) of light yellow crystal: mp 112 °C; MS m/e 266 (M+); ^{1}H-NMR δ ($CDCl_3$) 1.40(6H, bs, H-3~8), 2.06(4H, bs, H-1,2,9,10), 3.20~3.70 (4H, bs, amine protons), 6.55~6.60 (4H, d, H-12,12',13,13'), 7.00~7.03 (4H, d, H-11,11',14,14') ppm; ^{13}C-NMR δ ($CDCl_3$) 22.6(C-2,4), 26.0(C-3), 36.6(C-1,5), 43.8(C-6), 113.7(C-9,11), 127.0(C-8,12), 136.4(C-7) 145.5(C-10) ppm; FT-IR (KBr pellet) 3343, 2932, 1627, 1513, 1469, 1362, 1279, 1189 cm^{-1} ; Elemental Anal. Calcd. for $C_{18}H_{22}N_2$: C, 81.09 ; H, 8.26 ; N,10.51 , Found: C, 80.91 ; H,8.50 ; N, 10.24 %

All other diamines were prepared in the similar procedure to that of BACH with the corresponding cyclohexanone derivatives instead of cyclohexanone.

Polymer synthesis[11)]

Preparation of polyimides by chemical imidization. To a solution of BATM (5 mmol, 1.54 g) in 17.9 mL of NMP, BTDA (5 mmol, 1.61 g) was added with a solid content of 15 wt%. The solution was continuously stirred at room temperature for 24 h. To the reaction mixture were added 1.5 mL of acetic anhydride and 2.0 mL of pyridine and then the temperature was raised to 120 °C. After the reaction mixture was kept at 120 °C for 3 h, the resulting viscous solution was poured into excess methanol and filtered. The precipitated polymer was washed several times with water and methanol, and then the polymer was dried at 100 °C for 12 h in vacuo. The polymerization yield was almost quantitative.

Preparation of polyimides by one-step polymerization. To a solution of BATM (5 mmol, 1.54 g) in 19.0 mL of freshly distilled *m*-cresol, BTDA (5 mmol, 1.61 g) and isoquinoline(0.95 mL) as a catalyst were added at room temperature under nitrogen atmosphere. The reaction mixture was heated to 70~80 °C over 2 h and kept at that temperature for 2 h. And then the solution temperature was slowly raised to 200 °C over 2 h and refluxed for 6 h. The polymerization was performed under the gentle nitrogen stream to remove the water produced during imidization. The work-up process was same as the synthetic route (a).

Result and Discussions

Monomer Synthesis. The synthetic route of the monomers is shown in Scheme 1. Seven kinds of new aromatic diamines containing cyclohexylidene moiety with different alkyl-substituents were synthesized by a HCl-catalyzed condensation reaction of the corresponding cyclohexanone derivatives with excess aniline in an autoclave at temperature of 100~140 °C for 20~24 h. The yields of the monomer preparation were in the range of 32.0~83.1 %. Aromatic diamines prepared from cyclohexanone or 4-alkylcyclohexanone were obtained with comparably high yields above 70 ~ 83 %. However, the yield of compounds such as BATM, BANB and BAAD were only 30 ~ 40 %. It can be explained by the steric hindrance of corresponding cyclohexanone derivatives. The addition reaction of 1st aniline proceeded from the less hindered direction of cyclohexanone derivatives, which resulted in hydroxonium intermediate[11) . However, steric hindrance of cyclohexane ring retarded the substitution reaction of 2nd aniline. This showed that the dehydration reaction occurred faster than the substitution reaction of 2nd aniline in the sterically hindered cyclohexanone derivatives like 3,3,5-trimethyl cyclohexanone. The structures of prepared monomers were characterized by FT-IR, ^{1}H- and ^{13}C-NMR spectroscopy. The results of monomer synthesis are summarized in Tab. I.

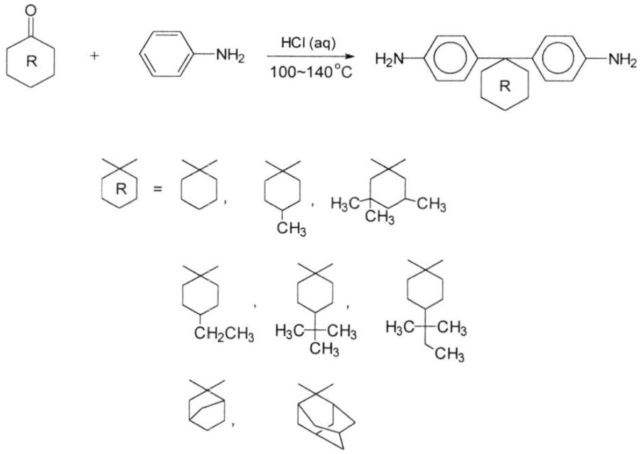

Scheme 1. Synthesis of monomers

Polymer Synthesis. For the synthesis of polyimides, two different synthetic routes were applied as shown in Scheme 2. At first, we had prepared the polyimides from BATM with aromatic dianhydrides by conventional polycondensation reaction followed by the chemical imidization using acetic anhydride and pyridine according to synthetic route (a). The inherent viscosities of the obtained polyimides were in the range of 0.55~0.68 dL/g, which might be due to the bended structure of BATM. However, one-step polymerization at high temperature as shown in synthetic route (b) could provide the enough energy even for the monomers with highly bent structure like BATM, which resulted in high viscosity as well as high molecular

Tab. I. Results of monomer synthesis

Monomer	Chemical Structures	m.p.(°C)	yield (%)
BACH	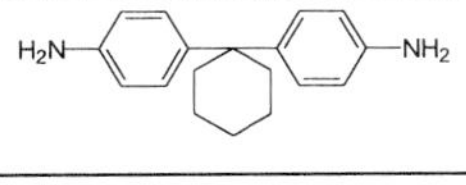	112.0	79.4
BAME	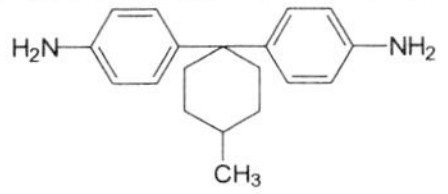	158.8	70.0
BAET	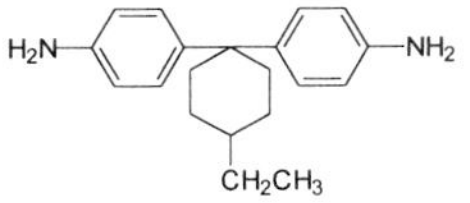	134.0	83.1
BABU	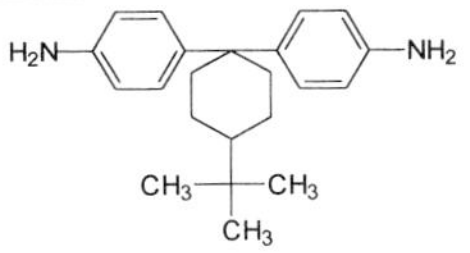	141.0	81.0
BAPE	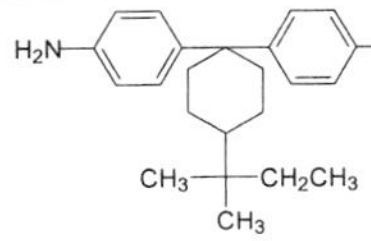	98.0	81.0
BATM	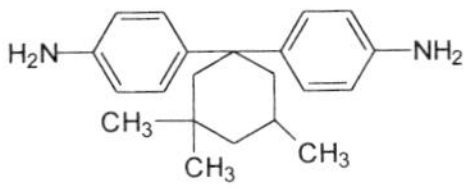	46.0	40.0
BANB	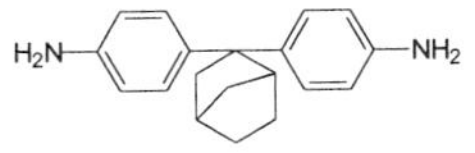	203	35.2
BAAD	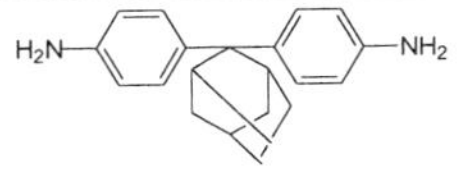	242	32.0

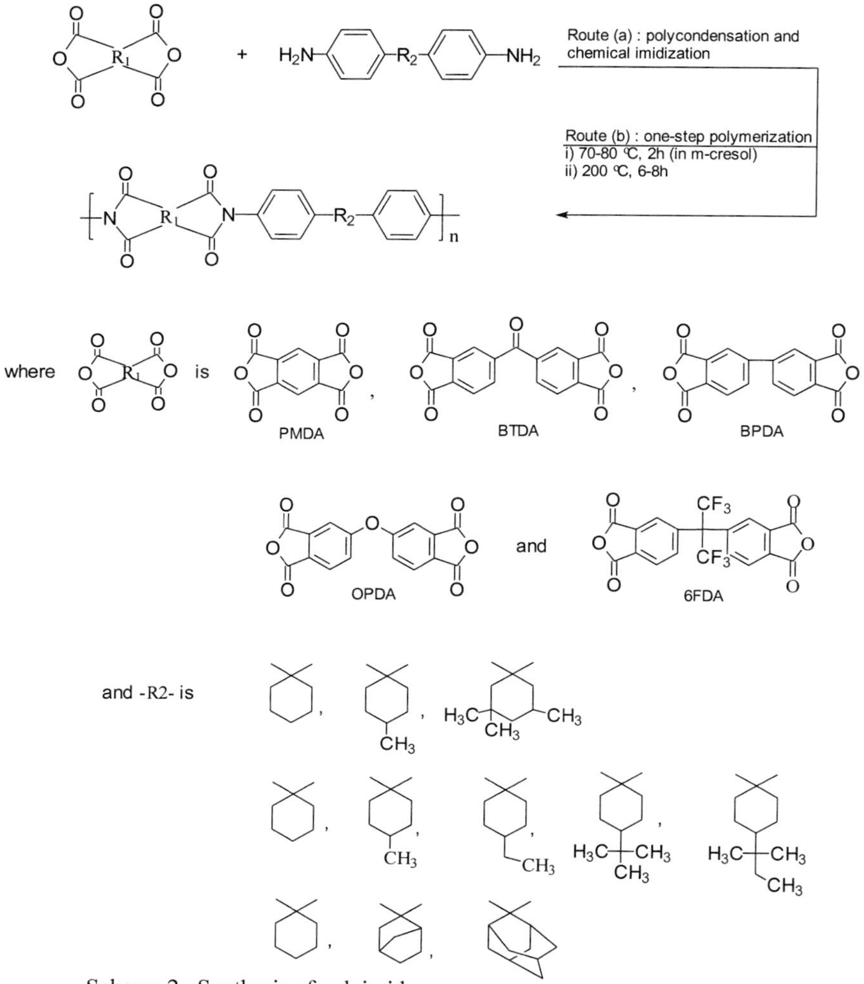

Scheme 2. Synthesis of polyimides

weight. The results of the polymerization are summarized in Tab. II. The inherent viscosities of polyimides obtained by one-step polymerization were in the range of 0.84~ 1.26 dL/g, which was remarkably increased as compared to those of polyimides prepared by chemically imidized two step polycondensation. Therefore, we have performed the one-step polyimidization in *m*-cresol at the temperature of 200 °C to obtain the polyimides with high molecular weights from the prepared diamines. Depending on the structure of dianhydrides, the solid contents were varied from 10 to 20 %. That is, the concentration had been kept low for the polymerization of PMDA with rigid structure to prevent the gel formation. The results of the polymerization are summarized in Tab. III. The inherent viscosities measured in *m*-cresol at 30 °C ranged from 0.37 to 1.66 dL/g. On the other hand, the relatively lower molecular weight of polyimides based on BACH resulted from their poor solubility to reaction medium. That is, polyimide(C-1) derived from BACH and PMDA was not soluble in *m*-cresol even at high temperature of 200 °C, which was precipitated out during polymerization without increasing molecular weight. The structures of the polyimides were characterized by FT-IR and ^{1}H- and ^{13}C-NMR spectroscopy. The FT-IR spectra of all polyimides exhibit the characteristic imide absorptions band in the regions of 1782, 1720 ($\nu_{C=O}$ asym. and sym. stretching, respectively), 1380-1350 ($\nu_{C\text{-}N}$), and 750-720 (imide ring deformation) cm^{-1}.

Tab. II. Synthesis of new soluble polyimides by different synthetic routes

Polymer	Diamines[a)]	Dianhydrides[b)]	Polymerization Method	η_{inh} (dL/g)	GPC Mn (g/mol)	GPC Mw (g/mol)	GPC PDI[f)]
T-1	BATM	PMDA	one-step[c)]	1.04	76,000[e)]	129,000	1.70
T-2	BATM	BTDA	one-step[c)]	1.21	87,000	133,000	1.53
T-3	BATM	BPDA	one-step[c)]	1.25	74,000	132,000	1.78
T-4	BATM	ODPA	one-step[c)]	1.26	92,000	159,000	1.97
T-5	BATM	6FDA	one-step[c)]	0.84	91,000	142,000	1.46
T-(1)	BATM	PMDA	two-step[d)]	0.68	60,000	109,000	1.82
T-(2)	BATM	BTDA	two-step[d)]	0.66	38,800	89,000	2.31
T-(3)	BATM	BPDA	two-step[d)]	0.56	40,000	94,000	2.32
T-(4)	BATM	ODPA	two-step[d)]	0.55	44,000	74,000	1.68
T-(5)	BATM	6FDA	two-step[d)]	0.55	60,000	102,000	1.70

[a)] Abbreviations : BATM; 1,-Bis(4-aminophenyl)-3,3,5-trimethylcyclohexane

[b)] Abbreviations : PMDA ; pyromellitic dianhydride, BTDA ; 3,3'4,4'-benzophenonetetracarboxylic dianhydride, BTDA; 3,3'4,4'-Biphenyltetracarboxylic dianhydride, ODPA; 3,3'4,4'-tetracarboxy-diphenyl oxide dianhydride, 6FDA;2,2-bis(3,4-dicarboxyphenyl)hexafluoropropane dianhydride

[c)] 5.0 mmole of diamine and dianhydride were reacted in 19.0 mL of *m*-cresol at 70~80 °C for 2 h and at 200 °C for 6~8 h under nitrogen atmosphere in the presence of isoquinoline as a catalyst.

[d)] Polymerization was performed with 5.0 mmol each of diamine and dianhydride in 10.5 mL of NMP at 20~25 °C for 24 h and followed by chemical imidization using Ac_2O/pyridine under nitrogen atmosphrene.

[e)]Measured in chloroform solution at 40 °C with μ-Styragel column HT6E5E3,

[f)]PDI; Poly Dispersity Index

Tab. III. Inherent viscosity (dL/g) of polyimides prepared by one-step polymerization[a]

Dianhydrides / Diamines	PMDA[b] (1)	BPDA (2)	BTDA (3)	6FDA (4)	ODPA (5)
BACH(C)[c]	-	0.93	1.03	0.57	0.66
BAME (M)	1.18	1.31	1.58	0.89	1.02
BAET (E)	1.27	1.26	1.23	0.92	1.57
BABU(B)	0.93	1.01	0.95	0.91	1.51
BAPE(P)	1.24	1.66	1.14	1.00	0.74
BATM(T)	1.04	1.25	1.21	0.84	1.26
BANB(N)	0.88	1.02	0.92	0.59	0.90
BAAD(A)	0.37	0.44	0.52	0.34	0.49

[a] 5.0mmole of diamine and dianhydride were reacted in 19.0 mL of *m*-cresol at 70~80 °C for 2 h and at 200 °C for 6~8 h under nitrogen atmosphere in the presence of isoquinoline as a catalyst.

[b] Abbreviations : PMDA ; pyromellitic dianhydride, BTDA ; 3,3'4,4'-benzophenonetetracarboxylic dianhydride, BTDA; 3,3'4,4'-Biphenyltetracarboxylic dianhydride, ODPA; 3,3'4,4'-tetracarboxy-diphenyl oxide dianhydride, 6FDA;2,2-bis(3,4-dicarboxyphenyl)hexafluoropropane dianhydride

[c] Abbreviations : BAME;1-1-Bis(4-aminophenyl)4-methyl cyclohexane, BAET;1,1-Bis(4-aminophenyl)-4-ethylcyclohexane, BABU; 1,1-Bis(4-amino phen- yl)-4-n-butylcyclohexane, BAPE; 1,1-Bis(4-aminophenyl)-4-n-pentylcyclohexane, BATM; 1,-Bis(4-aminophenyl)-3,3,5-trimethylcyclohexane, BANB, 2,2-bis(4-aminophenyl)norbornane; BAAD, 2,2-bis(4-aminophenyl) adamantane.

Characterization of polyimides

Solubility. All of the polyimides containing subsitituted cyclohexylidene moiety prepared in this study were readily soluble in typical aprotic dipolar solvent such as NMP, dimethylacetamide(DMAc), dimethylsulfoxide (DMSO) as well as 1,1,2,2-tetrachloroethane (TCE). That is, the introduction of alicyclic structure decreased the inter-chain interaction of rigid aromatic polyimides, which resulted in the improvement of solubility. Tab. IV compared the solublility of the polyimides based on the aromatic diamines containing the methyl substituted cyclohexane ring with that of polyimide without methyl substituent. Most of the polyimides derived from BACH which has no methyl substituent were not soluble in common organic solvents. On the other hand, polyimides obtained from BATM having three methyl substituents showed the best solubility. The three methyl substitutions might decrease inter-chain interaction of rigid aromatic repeat units, resulting in the improved solubility.

Furthermore, the polyimides prepared from diamines containing the alkyl-substituted cyclohexylidene moieties or polyalicyclic structures showed good solubility to common organic solvent like chloroform. As shown in Tab. V, the solubility of polymers was improved significantly by the introduction of alkyl groups as well as the polycycloalkane structures, which might be due to the bulky structure of diamines used. And, most of polyimides could be cast into flexible, tough, and colorless films from TCE solutions.

Tab. IV. Effect of the number of the methyl substituents on the solubility

Poly-imides	Dianhydride	Diamine	Solvents						
			NMP [a)]	DMSO	TCE	Dioxane	THF	$CHCl_3$	Acetone
C-1	PMDA	BACH	– – [b)]	– –	+ +	– –	– –	– –	– –
M-1	PMDA	BAME	+ +	+ +	+ +	+ +	– –	– –	– –
T-1	PMDA	BATM	+ +	+ +	+ +	+ +	+ +	+ +	– –

[a)] Abbreviations : NMP, *N*-methyl-2-pyrrolidone, DMSO; dimethylsulfoxide, TCE; 1,1,2,2-tetrachloroethane, THF; tetrahydrofuran

[b)] Key : + + : soluble; + : soluble on heating ; + – : partially soluble ; – – : insoluble

Tab. V. Effect of the size of alkyl substituents on the solubility

Poly-imides	Dianhydride	Diamine	Solvents						
			NMP [a)]	DMSO	TCE	Dioxane	THF	$CHCl_3$	Acetone
C-1	PMDA	BACH	– – [b)]	– –	+ +	– –	– –	– –	– –
M-1	PMDA	BAME	+ +	+	+ +	+ +	– –	– –	– –
E-1	PMDA	BAET	+ +	+	+ +	+ +	+ +	+ +	– –
B-1	PMDA	BABU	+ +	+	+ +	+ +	+ +	+ +	– –
P-1	PMDA	BAPE	+ +	+	+ +	+ +	+ +	+ +	– –

[a)] Abbreviations : NMP, *N*-methyl-2-pyrrolidone, DMSO; dimethylsulfoxide, TCE; 1,1,2,2-tetrachloroethane, THF; tetrahydrofuran

[b)] Key : + + : soluble; + : soluble on heating ; + – : partially soluble ; – – : insoluble

Thermal Properites. The glass transition temperatures (Tgs) of the polyimides measured by Differential Scanning Calorimetry(DSC) were ranged from 290 to 372 °C, which was higher than that of polyimide, C-1 prepared from diamine (BACH) with non-substituted cyclohexylidene moiety. That is, the Tgs were increased by the introduction of alkyl-substituents onto cyclohexane ring. As shown in Tab. VI, it was also found that Tgs were increased with the number of methyl substituents because of the increasing restriction on the

Tab. VI. Effect of the number of methyl substituents of polyimides on the glass transition temperature(°C)[a)]

Dianhydrides / Diamines	PMDA	BPDA	BTDA	6FDA	ODPA
BACH	-[b)]	-	305	293	290
BAME	349	355	346	306	300
BATM	-	~370(372)[c)]	336(347)	323(332)	330(316)

[a)]From the second heating traces of DSC measurements conducted with a heating rate of 10°C/min in nitrogen
[b)] Not detectable
[c)] Prepared by two-step chemical imidization

Tab. VII. Thermal properties of typical polyimides containing polyalicyclic structures

Polymers	Dianhydrides	Diamines	Tg(°C)[a)]	IDT(°C)[d)]	RW(%)[e)]
N-1	PMDA	BANB	-	430	47.4
N-2	BPDA	BANB	-[b)]	423	47.0
N-3	BTDA	BANB	347	384	47.3
N-4	6FDA	BANB	332	418	48.5
N-5	ODPA	BANB	- (327)[c)]	432	46.3
A-1	PMDA	BAAD	-	472	43.2
A-2	BPDA	BAAD	- (323)[c)]	463	46.8
A-3	BTDA	BAAD	363 (359)[c)]	396	48.6
A-4	6FDA	BAAD	349	394	49.3
A-5	ODPA	BAAD	331	423	46.0
C-1	PMDA	BACH	-	423	44.4
C-2	BPDA	BACH	-	441	44.1
C-3	BTDA	BACH	305	405	48.4
C-4	6FDA	BACH	293	436	43.5
C-5	ODPA	BACH	290	456	42.6

[a)] From the second heating traces of DSC measurements conducted with a heating a rate of 10 °C/min. under nitrogen atmosphere.
[b)] Not detected.
[c)] From MDSC(Modulated DSC) measurements conducted with a heating a rate of 2 °C/min. in nitrogen atmosphere.
[d)] IDT : Initial decomposition temperature measured by TGA at a heating rate of 10 °C/min under nitrogen atmosphere.
[e)] RW : residual weight at 800 °C

main chain rotational motion. Furthermore, as shown in Tab. VII, the Tgs of polyimides based on BAAD and BANB were higher than those of polyimides derived from BACH. That is, the incorporation of more rigid and bulky polycyclic structures restricted the free rotation of the polymer chain, which gave the higher Tgs. The thermal stabilities of polyimides were also investigated by thermogravimetric analysis(TGA) under nitrogen atmosphere.
The initial decomposition temperature of polyimides were more than 400 °C and the residual weight at 800 °C were above 30 %. Therefore, it can be concluded that the introduction of the cyclohexylidene moiety with alkyl-substituents along with the polyimide backbone improved the solubility with a retention of its high thermal stability.

Transmittance. UV-visible spectra of polymers based on BATM and various aromatic dianhydrides at a concentration of 5 x 10^{-3} mol/L in NMP solution were obtained . All the polymers showed high transmission above 90 % in the wavelength of 450~600 nm. That is, trimethylcyclohexylidene moiety increased the intermolecular chain distance and decreased the molecular interaction of the polymer chains, which resulted in good optical transparency. Especially, polyimides based on 6FDA exhibited better transparency than those of other polymer series. It can be seen that the electron conjugation of 6FDA moiety was also prohibited by the bulky trifluoromethyl structure. On the other hand, polyimides derived from PMDA showed strong absorption with a cut off between 330~ 450 nm, which was due to their electronic conjugation structure of PMDA.

Acknowledgement. We wish to acknowledge to Ministry of Science and Technology (MOST) of Korea for the financial support of this work with grant No KK-9804-06 .

References

[1] D. Wilson, H. D. Stenzenberger and P. M. Hergenrother, "*Polyimides*"; Blackie, Glasgow, 1990

[2] S. J. Huang and A. E. Hoyt, *TRIP*, **3**, 262 (1995)

[3] T. Matsumoto and T. Kurosaki , *Macromolecules*, **30**, 993 (1997)

[4] Y.Tsuda, Y. Tanaka, K. Kamata, N. Hiyoshi, S. Mataka, Y. Matsuki, M. Nishikawa, S. Kawamura ,and N. Bessho, *Polymer Journal*, **29**, 574 (1997)

[5] M. Kusama , T. Matsumoto, and T. Kurosaki, *Macromolecules*, **27**, 1117 (1994)

[6] W.Volkson, H. J. Cha, M. I. Sanchez, and D. Y. Yoon, *Reactive & Functional Polymers*, **30**, 61 (1996)

[7] Q. Jin, T. Yamashita, K. Horie, R. Yokota and I. Mita, *J. Polym. Sci., Part A: Polym. Chem.*, **31**, 2345 (1993)

[8] Q. Jin, T. Yamashita, and K. Horie, 3 , *J. Polym. Sci., Part A: Polym. Chem.*, **32**, 503 (1994)

[9] M. H. Yi , M. Y. Jin , and K.-Y. Choi, *Angew. Makromol. Chem.*, **233**, 89 (1995)

[10] K.-Y. Choi , M. H. Yi et al , *Korea Polymer Journal*. **4** , 117 (1996)

[11] M. H. Yi, W. Huang, M. Y. Jin, and K.-Y. Choi, *Macromolecules*, **30**, 19 (1997)

[12] M. H. Yi , W. Huang , J. T. Jung, S. K. Kwon, and K.-Y. Choi, *J. of Macromolecular Sci., A-Pure and Applied Chem.*, **A 35**(5), 821 (1998)

[13] M. H. Yi, W. Huang, B.J. Lee, and K.Y.Choi, *J. Polym. Sci.*, submitted.

[14] H. Waldmann, U. Leyrer, H. P.Mueller, K. J. Idel, C. Casser, G. Fengler, and U. Westeppe, *Ger. Offen DE* 4,014,847 (1991)

[15] J. Myska, *Chem. Abstr.* **61**, 14558 (1964)

Macromol. Symp. **142,** *205–215 (1999)*

Preparation and Photophysics of Polyimide/Silica Organic/Inorganic Hybrid Composites

Chang-Sik Ha* and Won-Jei Cho

Department of Polymer Science and Engineering, Pusan National University, Pusan 609-735, Korea

SUMMARY: In this work, polyimide/silica hybrid composites were prepared by the sol-gel reaction of tetraethoxysilane(TEOS) and the thermal imidization of poly(amic acid) from 3,3',4,4'-biphenyltetracarboxylic dianhydride(BPDA) and 4,4'-oxydianiline(ODA), and their photophysical properties were investigated using a fluorescence spectroscopy. It was found that the intrinsic fluorescence of poly(4,4'-oxydiphenylene-3,3'4,4'-biphenyltetracarboximide) (BPDA-ODA) such as emission intensity and emission wavelength depends strongly on the changes in the molecular conformations during the sol-gel reaction and the thermal imidization. In conclusion, we found that the fluorescence spectroscopy can provide an insight into how the intermolecular or intramolecular interaction of polyimide in the hybrid composite system is affected by the silica contents, depending on the sample states.

Introduction

The organic/inorganic hybrid composites constitute emerging advanced materials since they combine unique properties from the inorganic and organic components[1-4]. Recently, Nandi, et al[2] reported that aromatic polyimides can be a suitable matrix polymer for hybrid composites in advanced technological applications in the microelectronics and aircraft industries, since they possess high thermal stability, and excellent chemical, physical, and mechanical properties. In addition, silica has been most extensively investigated as an inorganic component for the hybrid composites because of the expected interesting catalytic and electronic applications[2].

Two methods of achieving the polyimide/silica hybrid composites are the direct addition of the oxide particles and the use of a precursor such as tetraalkoxysilane that produces the oxide particles during the second stage of thermal curing. The second approach, which may be accomplished with the sol-gel process, results very small particles with a more homogeneous distribution[3].

 CCC 1022-1360/99/$ 17.50+.50/0

Fluorescence spectroscopy is known to be very useful in the investigation of the dynamics and structure of solid polymers and is complementary to the X-ray methods[5]. Fluorescence spectroscopy has been widely used in the studies to reveal the molecular orientation of polyimide *per se* and to see the effects of curing conditions on the thermal imidization of poly(amic acid)s[5-11]. In the present study, therefore, we wish to learn how fluorescence spectroscopy can provide insight into whether the state of organization of the polyimide matrix is altered by the presence of silica or how the intermolecular or intramolecular interaction in the hybrid composite systems depends on the sample states.

Previous works on polyimide/silica hybrid composites have demonstrated that the properties of the polyimide matrix such as dielectric and tensile properties can be influenced by the presence of the silica[2),4]. Recently, we found that the polyimide morphology which had been investigated by atomic force microscopy and small angle X-ray scattering(SAXS) techniques is also significantly affected by the presence of silica and the types of the polyimide precursors[12].

Since the preparation of the polyimide/silica hybrid composites for this work includes the sol-gel reaction of tetraalkoxysilane, drying of the poly(amic acid) as-spun film, and thermal imidization, the molecular conformation of polyimide and the aggregate state of the inorganic silica particles as well as the dimension of silica networks are affected by several factors such as complexation between a solvent and poly(amic acid), chain orientation, chemical reaction, structural relaxation, and the interaction between silica and poly(amic acid) in solution or polyimide.

The photophysical behavior of the polyimide/silica hybrid composites will be therefore influenced by the presence of tetraalkoxysilane or silica gel in many ways depending on their interaction with poly(amic acid) in solution or poly(amic acid) as-spun film as well as polyimide.

We report here the effect of silica contents on the fluorescence behavior of the hybrid composite systems before and after sol-gel reaction of tetraethoxysilane(TEOS) to silica and/or thermal imidization to poly(4,4'-oxydiphenylene-3,3'4,4'-biphenyltetracarboximide) (BPDA-ODA) for which the emission intensity and emission wavelength depend strongly on the changes in the molecular conformations during sample preparation.

Experimental

Materials and Sample Preparation. 3,3',4,4'-biphenyltetracarboxylic dianhydirde (BPDA, TCI Chemicals) was recrystallized and vacuum-dried at 200°C for 24h before use. 4,4'-oxydianiline (ODA, Aldrich), and anhydrous N-methyl pyrrolidinone(NMP, Aldrich) were used as received. Tetraethylorthosilicate(TEOS) obtained from Aldrich and was used as received.

The BPDA-ODA poly(amic acid) (hereafter PAA, unless otherwise noted.) was prepared by adding an equimolar amount of BPDA powder into the NMP solution of ODA with continuous stirring at room temperature for several hours. The preparation of the BPDA-ODA polyimide-silica (hereafter PI-Silica, unless otherwise specified.) hybrid films is shown in scheme I. Various quantities of TEOS, water and HCl catalyst were then added into the poly(amic acid) solution (15wt.%). The weight ratio of TEOS to PAA varied from 0.05 to 1. The heterogeneous solution was stirred for 1 day until the solution became homogeneous. The resulting homogeneous mixtures were spun on glass plates or quartz wafers to prepare the PAA-Silica "as-spun" hybrid films. The PI-Silica hybrid films were obtained by successive heating the corresponding precursor PAA-Silica "as-spun" hybrid films. The PAA-Silica as-spun films after drying at 80°C for 30min were heated to 200°C for 30min followed by 350°C for 1h in a laminar flow forced air convection oven (Blue M) under nitrogen purge. Then the films were cooled to room temperature at a rate of 10°C/min. Fig. 1 shows the chemical structure of BPDA-ODA polyimide.

Measurement. The emission and excitation spectra of the films and solutions were measured at room temperature using a fluorescence spectrometer (FS 900CDT spectrometer, Edinburgh Instruments). The UV-Vis spectrometer (8452A diode array spectrometer, Hewlett Packard) was used for recording the absorption of films and solutions.

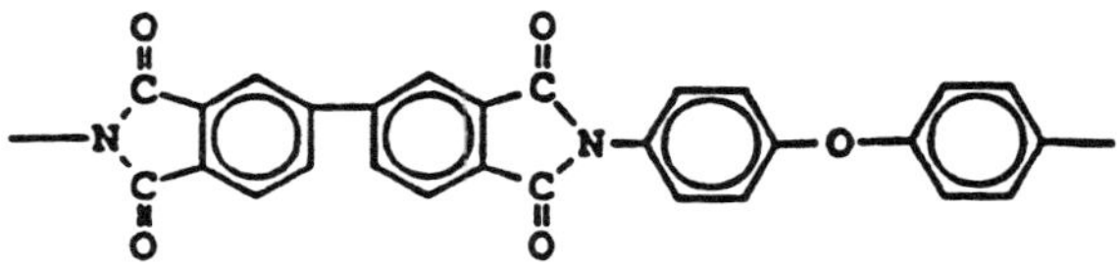

Fig. 1. Chemical structure of BPDA-ODA PI

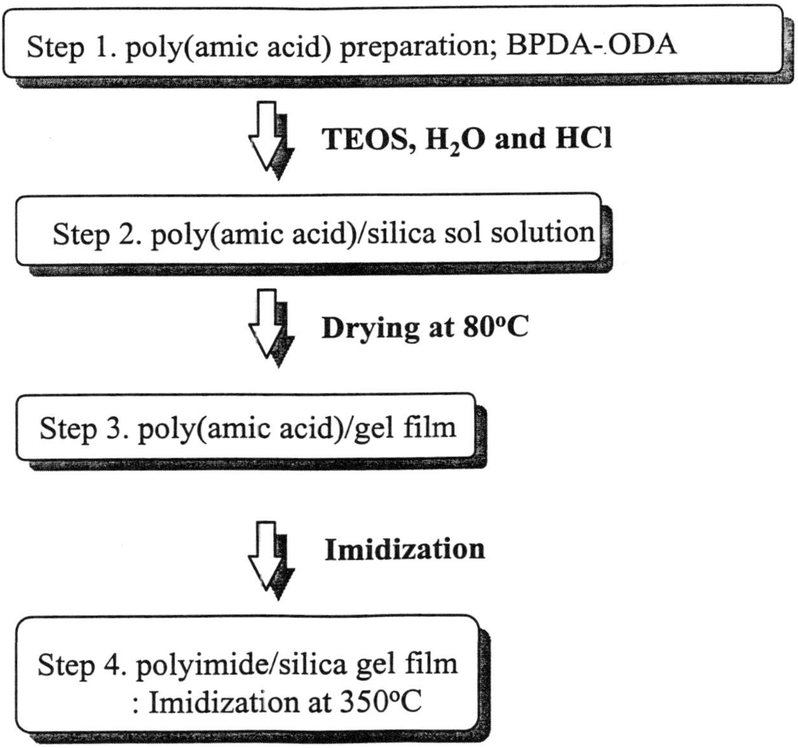

Scheme I: The preparation of polyimide-silica hybrid films

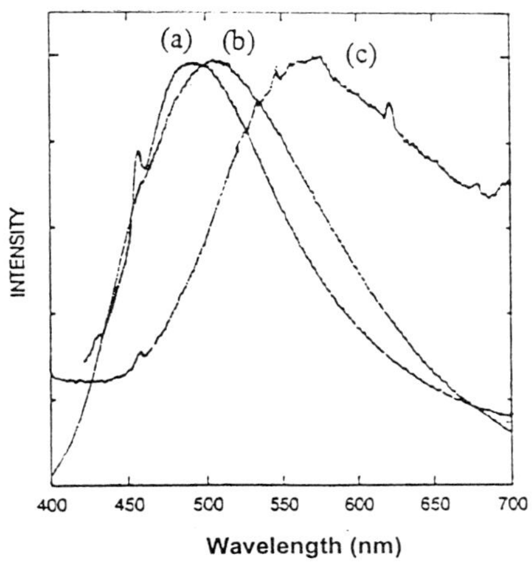

Fig.2. The emission spectra of the PAA solution in NMP (a), PAA as-spun film(b), and PI film(c), which were excited at 350 nm, one of the stronger excitation peaks in the excitation band, for PAA and PI films, and at 400 nm for PAA solution, respectively.

Results and Discussion

Fig. 2 illustrates the emission spectra of the PAA solution in NMP (a), PAA as-spun film(b), and PI film(c), which were excited at 350 nm, one of the stronger excitation peaks in the excitation band, for PAA and PI films, and at 400 nm for PAA solution, respectively. Broad structureless peaks were observed at approximately 505 nm, 509 nm and 570 nm for PAA solution, PAA as-spun film and PI films, respectively.

Even though no literature reports on the exact peak assignment of the chromophores in BPDA-ODA, we can deduce the nature of the chromophores in BPDA-ODA by reviewing literatures which deal with fluorescence of the polyimides or their model compounds either precursor of which is the same as BPDA-ODA such as poly(4,4'-

oxydiphenylenepyromellitimide) (PMDA-ODA), poly(4,4'-oxydiphenylene-3,3'4,4'-benzophenonetetracarboximide) (BTDA-ODA), and poly(1,4-phenylene-3,3'4,4'-biphenyltetracarboximide) (BPDA-PDA). Previous workers have proposed that intermolecular and/or intramolecular charge transfer complexes (CTC) exist and give rise to the fluorescence in those polyimides[5-11)]. The intensity of the CTC depends not only on the nature of chromophores but also on the thermal imidization conditions.

For the PI film, broad and structureless excitation peaks, measured with 550 nm emission, were observed around 350 and 480 nm, while broad emission peak of the PI film was observed around 570 nm for both short-wavelength (350 nm) and long-wavelength(480 nm) excitation. Similar excitation and emission spectra were also reported by Hasegawa, et al[6)]. It is generally accepted that polyimide fluorescence usually arises from CTC's that form between electron donor (diamine) and electron acceptor (diimide) segment pairs[7), 11)]. It is believed for the BPDA-ODA PI the emission peak is due to the intermolecular CTC[11)]. We note that the emission peaks of PI films were observed at much longer wavelength than those of PAA as-spun films as well as PAA solution. The change of fluorescence of the PAA as-spun film to PI film is of course due to the structural development such as in-plane orientation and molecular packing[6-11)]. Hasegawa et al.[8)] reported that the changes in the fluorescence intensity are due to the energy transfer from PAA units to PI units as imidization proceeds. Wachman and Frank[11)] reported that the fluorescence is a measure of some aspect of the molecular rearrangement between the polyimide chains depending on the curing temperature. Bessonov, et al.[13)] reported that the interaction between the highest occupied molecular orbital(HOMO) of the amine, bearing a lone electron pair, with the lowest unoccupied molecular orbital(LUMO) of the dianhydride is very important for the imidization reaction since they are closest in energies.

It was reported by the authors[14)] that in case of BPDA-PDA(p-phenylenediamine) the emission peak is observed at much longer wavelength for BPDA-PDA PAA solution than is for BPDA-PDA as-spun film as well as BPDA-PDA PI film, meaning the formation of tight complexation between solvent, NMP, and PAA. However, such difference in the emission peak wavelength was not clearly observed for BPDA-ODA. The result indicates that the complex between BPDA-ODA PAA and NMP is not tight.

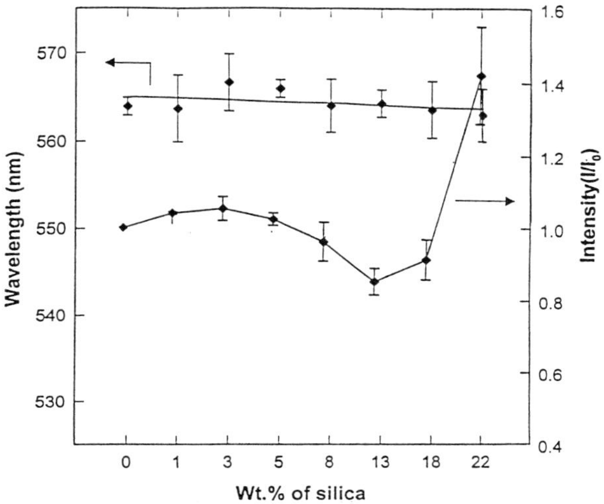

Fig. 3. The effect of silica contents on the emission wavelength and intensity of PI/Silica composite films. The content of silica was calculated assuming that the gel reaction from TEOS to silica was completed.

Brekner and Feger[15)] reported that imidization is hindered by the complexation, if any. They reported that the hydrogen-bonded complex formed between the solvent NMP and PAA can prevent the reactive groups of the PAA from getting into the conformation required for imidization to occur.

The decomplexation of amic acid from NMP proceeds when the PAA film is exposed at higher temperature, because of decreasing solvent diffusion with degree of imidization. It may be assumed that when the PAA film was soft-baked at 80 °C, several structural changes occur simultaneously and competitively including complexation and/or decomplexation between NMP and amic acid, and molecular ordering of amic acid to imide depending on the degree of imidization as well as solvent removal rate.

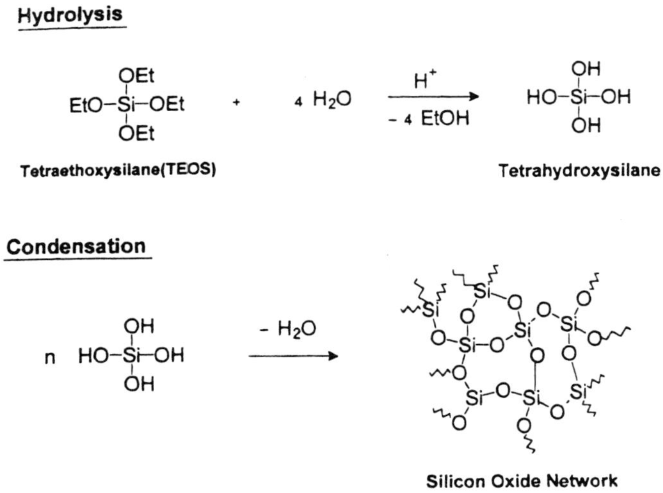

Fig.4. Sol-gel process of TEOS. Hydrolysis and condensation occur simultaneously[12].

Fig. 3 shows the effect of silica contents on the emission wavelength and intensity of PI/Silica composite films. In this figure, the wt. % of silica was calculated assuming that the gel reaction from TEOS to silica was completed. The error bar is the range of intensities for seven measurements. On increasing silica content, the wavelength was not much changed but the intensity slightly decreased except at high silica content.

The results of Fig. 3 suggests that the interaction between PI and silica particles is not strong. The result implies that the intermolecular and/or intramolecular charge transfer from HOMO energy level of ODA to the LUMO of imide (BPDA) is strong enough to hinder the interfacial interaction between the PI and the electron accepting silica. It was reported by the authors[14], however, that the emission maximum of the BPDA-PDA polyimide film was red-shifted with increasing silica contents, meaning that interfacial interaction exists between the PI and silica. The decreasing behavior of the emission intensity with increasing silica contents is due to the reduced population of charge transfer complex in the PI because of the nonfluorescent silica particles.

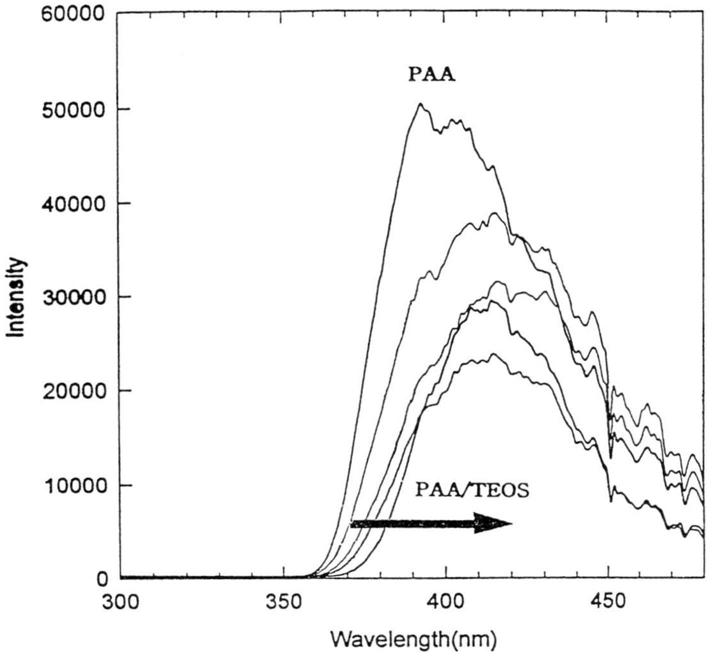

Fig. 5. The excitation spectra of the PAA/TEOS solutions with different TEOS contents. The arrow indicates the increasing TEOS contents. The emission was monitored at 500 nm.

Too many silica particles exhibit some scattering effect because of poor interaction between the PI and silica particles. In Fig.3, the very high emission intensity of the PI/Silica composite film, when the silica content is highest (about 22% of silica), is due to scattering from silica particles.

TEOS is readily hydrolyzed to $Si(OH)_4$ by water in the presence of an acid catalyst such as HCl, as shown in Fig. 4. The hydrolysis reaction is very fast even in the NMP solution of TEOS and PAA. Then the intermolecular hydrogen bonding exists between the hydroxy group in the hydrolyzed TEOS and carboxylic acid in BPDA in PAA systems.

Fig. 5 shows the excitation spectra of the PAA/TEOS solutions with different TEOS contents. In this figure, the emission was monitored at 500 nm. One can see that regardless of the TEOS contents, the wavelength of the excitation peak of the PAA/TEOS solution is red-shifted to the almost same wavelength, when compared to the PAA solution without a catalyst, HCl as well as TEOS.

Since TEOS is readily changed to $Si(OH)_4$ in the presence of a catalyst, the red-shift in the excitation peak is caused by the intermolecular hydrogen bonding between the hydrolyzed TEOS, *i.e.* $Si(OH)_4$ and PAA, regardless of whether the hydrolysis is completely or partially done. This may be true although TEOS is not completely hydrolyzed, since the same experimental results are obtained for different pH solutions by using other acid or base catalysts such as formic acid and t-ethylamine. Details on this point are described elsewhere[14].

Conclusion

The fluorescence spectroscopic analysis in this work showed that in BPDA-ODA PAA/TEOS solutions, the complexation between NMP and PAA is not tight but the hydrogen bonding between $Si(OH)_4$ and PAA is strong. When PAA solution is imidized to PI films, however, the interfacial interaction between PAA and silica is not strong, though decomplexation takes place during the imidization process and the system has no complexation effect.

In summary, we have observed some interesting fluorescence behavior to give an insight on the interfacial interaction in the polyimide/silica hybrid film, depending on the sample state, whether it is PAA solution, PAA as-spun film or PI film. Further fluorescence studies are now under way to provide more details on the interfacial interaction between poly(amic acid) or polyimide and silica as well as the morphology of the systems, and the effect of silica on the state of organization of the polyimide matrix.

Acknowledgments. C.S.H. wishes to thank Pusan National University (PNU), Korea for sabbatical leave and Stanford University, U.S.A. for kind hospitality during the leave. This work was supported by the Overseas Research Grant of the Seoam Scholarship

Foundation (Korea). He also wishes to express his sincerest thanks to Prof. Curtis W. Frank of Dept. of Chemical Engineering, Stanford University for his helpful discussion on this work. The experimental aid of Mr. Hae-Dong Park should be appreciated.

References

[1] S. Wang, A.Ahmad, J.E. Mark,. *Chem.Mater.* **6**, 943 (1994).

[2] M. Nandi, J.I. Conklin, J.L. Salvati, A. Sen, *Chem.Mater.* **3**, 201 (1991).

[3] P. Repaci , C.W. Frank, unpublished results.

[4] Y. Kim, M. Ree, T.H. Chang, E.J.Kang, C. Chang, M. Chang, W.J. Cho, C.S. Ha, *Synth.Metal.* **85**, 1399 (1997).

[5] M. Hasegawa, M.Kochi, A.Mita, R.Yokota, *Eur.Polym.J.* **25**,4,349 (1989).

[6] M. Hasegawa, I.Mita, M. Kochi, R.Yokota, *J.Polym.Sci.,Part C: Polym.Letters* **27**,26 (1989).

[7] M. Hasegawa, A. Arai, I. Mita, R.Yokota, *Polym. J.* **22**,10, 875 (1990).

[8] P.S. Martin, E.D. Wachsman, C.W. Frank, *"Polyimides:Materials, Chemistry and Characterization"*, C. Feger, M.M.,Khojasteh, J.E.MaGrath, Eds. Elsevier Science Publ. Amsterdam, 1989, pp.371-378.

[9] D.A. Hoffmann, H.Ansari, C.W.Frank, *Mat. Res. Soc. Symp. Proc.* **227**, 131(1991).

[10] J.M..Salley, C.W. Frank, *Charge Transfer in Aromatic Polyimides*, in M.K. Ghosh, Mittal, K.L. Eds., "*Polyimides*", Dekker, New York, 1996, pp.279-307.

[11] E.D. Wachman, C.W.Frank, *Polymer* **29**,1191 (1988).

[12] Y.Kim, W.J. Cho, C.S. Ha, M. Ree, T.H.Chang, *Polym. Int'l.* **43**,129 (1997).

[13] M.I. Bessonov, M.M. Koton, V.V. Kudryavtsev, C.A. Laius, "*Polyimide, Thermally Stable Polymers"*, Consultant Bureau, New York , 1987, p.18.

[14] C.S.Ha, H.D. Park, C.W. Frank, *Chem. Mater.*, submitted.

[15] M.J. Brekner, C. Feger, *J.Polym.Sci. Part A: Polym. Chem.* **25** ,2479(1987)).

Macromol. Symp. **142,** 217–225 (1999)

The Volume Transition in Thermosensitive Core-shell Latex Particles Investigated by Small-angle X-ray Scattering and Dynamic Light Scattering

J. Kim, I. Deike, N. Dingenouts, Ch. Norhausen, M. Ballauff*

Polymer-Institut, Universität Karlsruhe, Kaiserstr. 12, 76128 Karlsruhe, Germany

Summary: We present a survey over recent studies of the volume transition in colloidal core-shell particles composed of a solid poly(styrene) core and a shell of a thermosensitive crosslinked polymer chains. The thermosensitive shell is built up from poly(N-isopropylacrylamide) chains (PNIPA) crosslinked by N,N'-methylenbisacrylamide (BIS). In addition, particles containing acrylic acid (AA) as comonomer have been synthesized and investigated. The volume transition of these particles have been studied by dynamic light scattering (DLS) and by small-angle X-ray scattering (SAXS). In all cases analyzed so far the volume transition was found to be continuous. This finding shows that the core-shell microgels behave in a distinctively different manner than ordinary thermosensitive gels: The crosslinked chains in the shell are bound to a solid boundary independent of temperature. The spatial constraint by this boundary decreases the maximum degree of swelling but also prevents a full collapse of the network above the volume transition.

Introduction

Gels composed of cross-linked poly(N-isopropylacrylamide) (PNIPA) chains may undergo a volume transition as a function of temperature in which the network shrinks in a continuous or discontinuous fashion[1)]. Hence, gels composed of PNIPA and suitable comonomers may be shrunken by external stimuli which has lead to a number of interesting applications[2)]. Tanaka and coworkers demonstrated that incorporation of charged comonomers as e.g. acrylic acid increases the degree of swelling dramatically[1,3)]. In particular, depending on the content of charged co-units the continuous volume transition may become a discontinuous process in which a highly swollen gel shrinks at a given temperature to a dense state.

The slow dynamics of this process in macroscopic gels has lead to the synthesis of microgel particles in which the response to external stimuli will be much faster than in macroscopic systems[1)]. Hence, gel beads with diameters in the sub-millimeter range could serve for the elucidation of the swelling kinetics of macroscopic systems[4]. Recent experiments conducted

 CCC 1022-1360/99/$ 17.50+.50/0

by Kratz and Eimer[5)] on PNIPA-microgels having colloidal dimensions suggest that these systems behave in a similar way as macroscopic gels[6, 7)].

If a thermosensitive network is affixed on the surface of colloidal particles, however, the spatial constraint effected by this geometry is expected to exert a marked influence on the phase transition. Such core-shell microparticles have been prepared by a number of groups[8-12)]. In a recent paper the volume transition in a shell of ca. 30nm thickness on a poly(styrene) (PS) particle has been studied by small-angle X-ray scattering (SAXS)[13)]. Here it could be demonstrated that fixing the chains to a solid particle leads indeed to a profound change of the volume transitions. Rheological studies by Senff et al.[14)] furthermore showed that suspensions of these particles present ideal model systems for the investigation of the flow behavior of sterically stabilized particles.

In this paper a survey over recent studies on these thermosensitive core-shell particles will be given. Special emphasis is laid on the alterations effected on the volume transition by the presence of charged groups in the thermosensitive network of the shell.

Uncharged networks

Recently, we have presented the synthesis and the characterization of latex particles having a PS-core of ca. 80nm diameter onto which a shell of crosslinked PNIPA-chains has been affixed[13)]. The volume transition in these particles has been studied by dynamic light scattering and by SAXS. Here it could be shown that SAXS is uniquely suited to investigate the swelling and deswelling of the shell upon change of temperature. Figure 1 displays schematically the structure of such particles and their analysis by SAXS[13)]. Here are three terms which contribute to the observed scattering intensity I(q) ($q=(4\pi/\lambda)\sin(\theta/2)$; θ: scattering angle; λ: wave length)[13)]:

$$I(q) = I_{CS}(q) + I_{network}(q) + I_{PS}(q) \qquad (1)$$

Here $I_{CS}(q)$ is the part of I(q) due to the core-shell structure of the particles, i.e., the scattering intensity caused by a composite particles having a homogeneous core and a shell. The core and the shell are characterized by a different electron density which is shown schematically in fig. 1. The shell consists of a polymeric network which exhibits static inhomogeneities and thermal fluctuations taken into account by $I_{network}(q)$ (see the discussion of $I_{network}(q)$ in ref.[16, 17)]). Finally, $I_{PS}(q)$ denotes the scattering intensity which is caused by the density fluctuations of the solid PS-core of the particles.

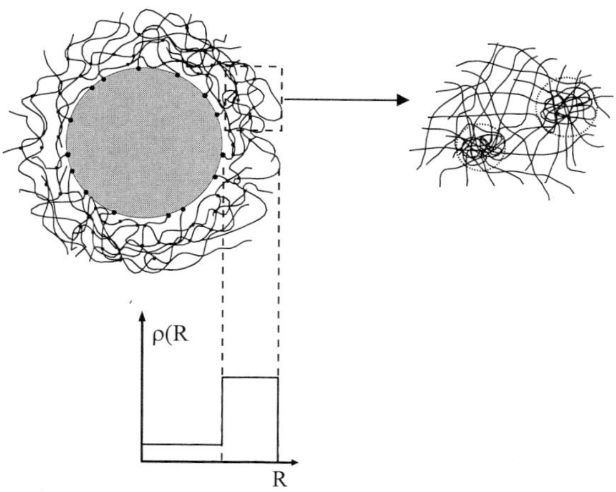

Fig. 1 Analysis of core-shell microgels by SAXS. The term $I_{CS}(q)$ (see eq.(1)) which is dominant at low q-values refers to a core-shell particle with the excess electron density ρ(R). The network exhibits static and dynamic inhomogeneities which lead to $I_{network}(q)$. Finally, the poly(styrene) cores of the particles are not entirely homogeneous but the density of the solid poly(styrene) fluctuates weakly around a mean value. This gives the third term $I_{PS}(q)$ which gives a small but non-negligible contribution to I(q). All three contributions to I(q) derive from statistically independent fluctuations of the electron density in the system and no cross term appears in eq.(1).

The analysis of these terms[13)] show that $I_{PS}(q)$ can easily be determined from the SAXS-intensity of the core particles. Furthermore, $I_{CS}(q)$ decreases approximately with q^{-4} whereas $I_{network}(q)$ diminishes with q^{-2}. Experimental data corroborate this prediction: Figure 2 displays I(q) obtained at 25°C (squares) and at 50°C (triangles). For the sake of clarity the latter curve has been multiplied by 100.

The decomposition of I(q) may be done as follows: First $I_{PS}(q)$ obtained from the core-latex (see above) is subtracted from the measured intensity I(q). In a second step $I_{CS}(q)$ is determined from I(q) taking into account only the data in the q-range smaller than $0.2 nm^{-1}$. The lines in fig. 2 show the respective fits achieved by the fit routine described recently[18,13)]. The excess electron densities $\rho(R)-\rho_m$ resulting from this fit are displayed in the inset of fig. 2 where ρ_m is the electron density of water. The profiles $\Delta\rho(R) = \rho(R) - \rho_m$ shown in the inset of fig. 2 indicate the swelling of the PNIPA-shell when going from poor solvent conditions (50°C; dashed lines) to good solvent conditions (25°C; solid lines). Furthermore, it can be shown that virtually all NIPA used in the synthesis is built into the thermosensitive shell.

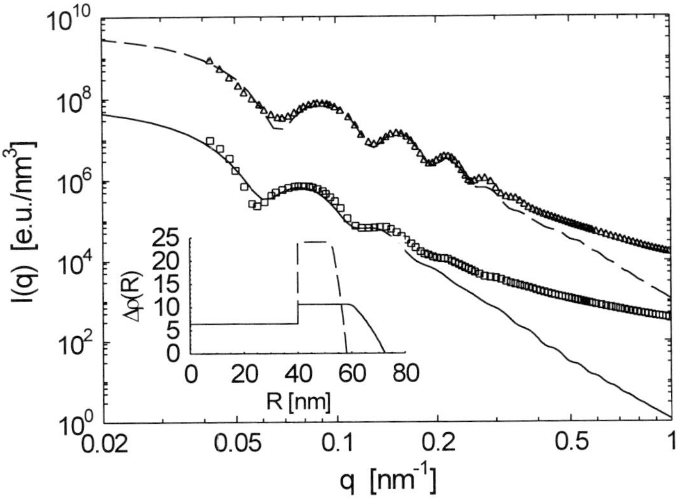

Fig. 2 Measured SAXS-intensities I(q) of the core-shell particles below the volume transition (25°C; squares) and above the volume transition (50°C; triangles). For the sake of clarity the data measured at 50°C have been multiplied by 100. The lines display the fit of $I_{CS}(q)$ (cf. eq.(1)) to the data at low q-values ($q<0.2nm^{-1}$; 25°C: solid line; 50°C: dashed line). The radial excess electron density $\Delta\rho(R)$ deriving from this fit is shown in the inset. The marked deviation between I(q) and I_{CS} (q) seen at high q-values is due to $I_{network}(q)$, i.e., the contribution of the network in the shell of the particles.[13)]

From the profile shown in the inset in fig. 2 the average volume fraction ϕ of PNIPA can be estimated to be 0.5 at 50°C whereas a value of 0.23 follows from the SAXS-analysis at 25°C. At 25°C the shell has an extension of ca. 32nm which shrinks to 18nm above the transition. This corresponds to a shrinking of the volume of the shell by a factor of 2.2 which is small compared to PNIPA macrogels having a similar degree of crosslinking[6, 7)]. It must be noted that the gel does not collapse but remains swollen to a certain extend even far above the volume transition.

This point can be discussed further when analyzing the excess scattering intensity visible at higher q-values in fig. 2. Here the difference between the measured intensity I(q) corrected for $I_{PS}(q)$ and $I_{CS}(q)$ can be assigned to $I_{network}(q)$ and analyzed accordingly[13)]. The main contribution to the latter term is deriving from the thermal fluctuations of the network[16)] which may be described by a Lorentzian[17)]:

$$I_{network}(q) \approx I_{fluct}(q) = \frac{I_{fluct}(0)}{1+\xi^2 q^2} \qquad (2)$$

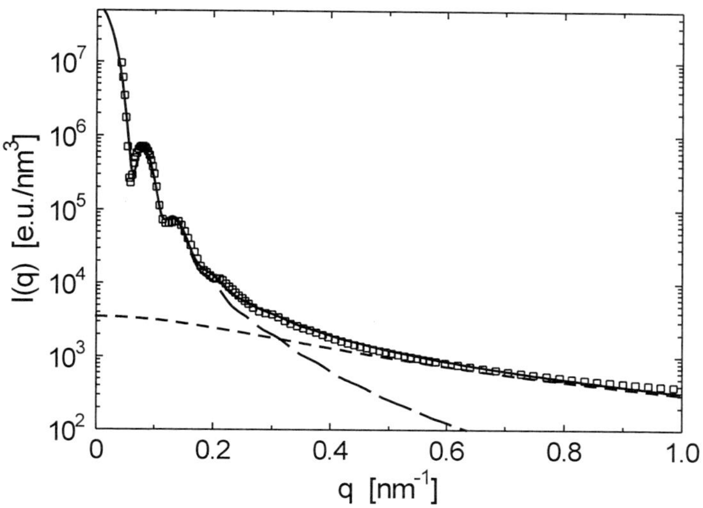

Fig. 3 Comparison of $I_{CS}(q)$ (long-dashed line) and $I_{fluct}(q)$ (short dashed line) with the scattering intensity $I(q)$-$I_{PS}(q)$ measured at 25°C (squares). The solid line gives $I_{CS}(q)$+ $I_{fluct}(q)$ (cf. eq.(1) and (2))[13].

where ξ denotes the correlation length of the thermal fluctuations. As demonstrated by figure 3, the measured scattering intensity $I(q) - I_{PS}(q)$ (cf. eq.(1)) can be described quantitatively by the superposition of $I_{CS}(q)$ dominating at small q and of $I_{fluct}(q)$ dominating at high q-values.

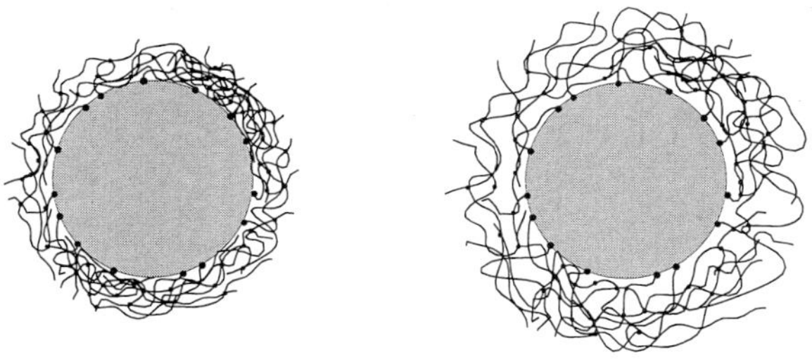

	poor solvent conditions 50 °C	good solvent conditions 25 °C
d_{shell}	18nm	32nm
$\phi_{network}$	0.50	0.25
ξ	2.1nm	3.2nm

Fig. 4 Analysis of the volume transition in uncharged thermosensitive core-shell particles[13].

Fig. 4 displays the main results of the SAXS-analysis of the uncharged core-shell particles: Due to the fixation of the chains of the network on the core particles the volume transition is strongly restricted, i.e., the transition does not lead to a fully collapsed state but only to a markedly higher volume fraction indicated in fig. 4. The chains can only shrink along the direction of the surface normal but no marked contraction can result parallel to the surface. If the unconstrained gel shrinks by a factor of 10, the corresponding shrinking of the microgel layer is expected to be of the order $(10)^{1/3} \approx 2.15$ which is of the order observed in the present system (cf. fig. 4).

Charged networks

The crosslink density of the PNIPA-network in the thermosensitive core-shell particles discussed so far was quite high and the volume transition was expected to proceed in a continuous fashion[1, 3)]. It is thus interesting to modify the network by charged groups to see whether the volume transition in the shell will become discontinuous as is the case for macroscopic networks[1, 3)]. Here we present first results obtained by dynamic light scattering (DLS) from core-shell particles in which the thermosensitive network of the shell has been modified by incorporation of acrylic acid (AA). As in the previous study[13)] crosslinking has been achieved by use of N,N'-methylenebisacrylamide (BIS; 5Mol% with regard to NIPA). Starting from narrowly distributed PS-core particles the shell is produced by the method described recently[13)]. For comparison one core-shell system has been prepared without AA-units. Details of the synthesis and characterization of the particles will be reported elsewhere[15)].

Fig. 5 displays the hydrodynamic radii R_H obtained by DLS of an uncharged system and of a core-shell system containing 5Mol% AA. This comparison demonstrates that the incorporation of a small amount of charged units suffices to increase the degree of swelling enormously. While R_H of the uncharged particles increases gradually to ca. 120nm at 10^0C, the charged system is characterized by a hydrodynamic radius of ca. 200nm despite the fact that the degree of crosslinking is the same. It is also interesting to note that the network in the shrunken state has approximately the same size as has been found for the uncharged particles; R_H of the charged and uncharged particles virtually coincide above 40^0C.

The data shown in fig. 5 furthermore show clearly that the transition seen in R_H is fully reversible; within experimental uncertainty (3%) the same hydrodynamic radius is obtained in the heating as well as in the cooling run. As in the case of uncharged particles the colloidal

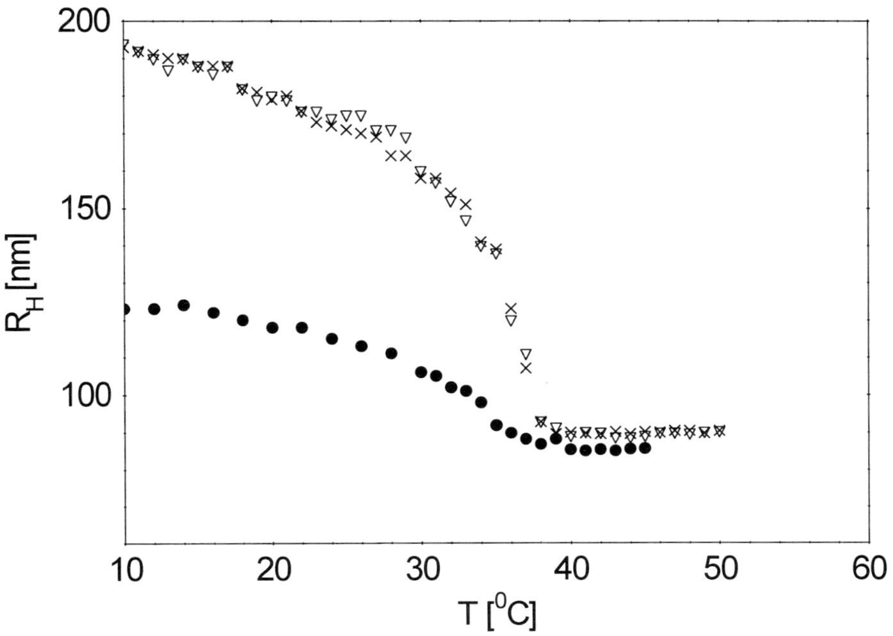

Fig. 5 Hydrodynamic radius R_H as function of temperature for thermosensitive core-shell networks without charges (filled circles) and with 5mol% acrylic acid co-units (triangles and crosses). All measurements have been done in pure water. Crosses denote the data obtained on cooling whereas triangles display R_H obtained from heating runs[15].

stability of the system may be traced back to the restricted shrinking of the network during the volume transition. The chains in the network are retained in a fluid-like state despite the fact that the network has been shrunken considerably. Hence, the steric stabilization of the particles will still be operative above the volume transition. In addition to this, residual charges on the surface of the core particles introduced by the initiator will enhance the stability of the particles above the volume transition.

A further important point seen in fig. 5 commands attention: Despite the charges in the network the volume transition as detected by R_H is continuous. The transition is taking place in a temperature range of ca. 5^{0}C and not within less than one degree as expected for a discontinuous transition (cf. e.g. fig. 5 of ref.[5)]). From the complete reversibility of the data it is obvious that the continuous variation of R_H is not due to coagulation. Also, the size distribution of the particles is rather narrow (standard deviation: 3%) so that no smearing out of the transition by polydispersity of sizes and structure may take place.

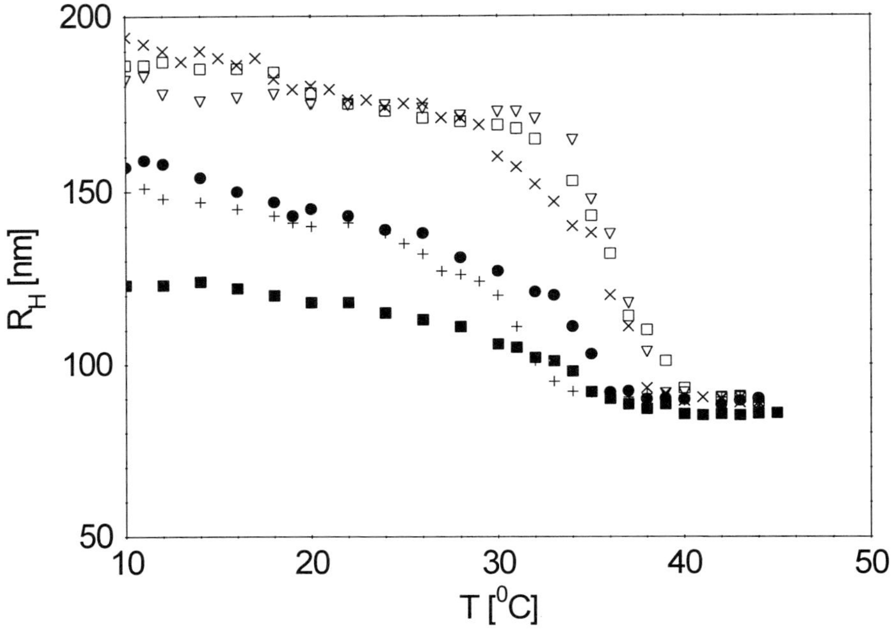

Fig. 6 Hydrodynamic radius R_H as a function of temperature for thermosensitive core-shell networks containing 5Mol% acrylic acid co-units at different pH. ■: uncharged core-shell particles CS;: +: charged particle at pH= 2; ●: pH = 4; x: pH = 7; □: pH=9;∇: pH = 12.

In must be kept in mind that the number of charges in the network is large enough to cause a discontinuous transition in macroscopic system[1,3]. The results displayed in fig. 5 clearly demonstrate, however, that the volume transition in a thin shell of colloidal particles proceeds in a continuous fashion.

The influence of the charges in the network can be tuned by the pH-value of the solution[15]. Figure 6 displays the hydrodynamic radius of the charged core-shell networks measured at different pH. The pH has been adjusted by adding small amounts of NaOH to dilute systems in order to avoid a concomitant change of ionic strength. Lowering the pH below the pK_s (5.6) of poly(acrylic acid) the degree of swelling decreases but remains above the respective value of the uncharged system. Above the pK_s of PAA the degree of swelling is markedly increased and nearly independent of the pH. Here all charges are ionized leading to the maximum degree of swelling. Again it becomes obvious that the variation of R_H remains continuous for all pH-values despite the fact that the degree of swelling varies strongly at different pH.

We ascribe this finding to the spatial constraint exerted by the surface of the core particles onto which the chains of the network are affixed. Together with the results[13)] obtained from uncharged networks it is obvious that core-shell micronetworks present an interesting class of gels in which the chains are subjected to geometric constraint. These constraints exerted by the solid surface of the core particles lead to a continuous volume transition even when in case of charged networks which undergo a discontinuous transition in macrogels.

Acknowledgment: Financial support by the Bundesministerium für Forschung und Technologie, and by the Deutsche Forschungsgemeinschaft, and by the Deutsche Forschungsgemeinschaft is gratefully acknowledged.

Literature

1. M. Shibayama, T. Tanaka, T. *Adv. Polym. Sci.*, **109**, 1 (1993); see also further reviews in this volume and volume 110
2. Y. Osada, J. P. Gong, *Adv. Mater.* **10**, 827 (1998)
3. T. Tanaka, D. Fillmore, S.-T. Sun, I. Nishio, *Phys. Rev. Lett.* **45**, 1636 (1980)
4. T. Tanaka, E. Sato, Y. Hirokawa, S. Hirotsu, J. Peetermans, *Phys. Rev. Lett.* **55**, 2455 (1985)
5. K. Kratz, W. Eimer, *Ber. Bunsenges. Phys. Chem.* **102**, 848 (1898)
6. S. Hirotsu, *Adv. Polym. Sci.* **110**, 1 (1993)
7. Y. Li, T. Tanaka, *J. Chem. Phys.* **90,** 5161 (1989)
8. K. Makino, S. Yamamoto, K. Fujimoto, H. Kawaguchi, H. Oshima, *J. Colloid Interf. Sci.* **166**, 251 (1994)
9. M. Okubo, H. Ahmad, *Colloid Polym. Sci.* **274**, 112. (1996)
10. Y. Nakazawa, Y. Kamijo, K. Fujimoto, H. Kawaguchi, Y. Yuguchi, H. Urakawa, K. Kajiwara, *Angew Makromol Chem* **240**,187 (1996
11. D. Duracher, F. Sauzedde, A. Elaissari, A. Perrin, C. Pichot, *Colloid Polym. Sci.* **276**, 219 (1998)
12. H. Kawaguchi, H. Matsuoka, K. Fujimoto, *Abstracts of the World Polymer Congress* 1998
13. N. Dingenouts, Ch. Norhausen, M. Ballauff, *Macromolecules* **31**, 8912 (1998)
14. H. Senff, W. Richtering, Ch, Norhausen, A. Weiss, M. Ballauff, *Langmuir* **15**, 102 (1999)
15. J. Kim, M. Ballauff, in preparation
16. M. Shibayama, *Macromol. Chem. Phys.* **199**, 1 (1998)
17. M. Shibayama, T. Tanaka, C. Han, *J. Chem. Phys. 97*, 6829 (1992)
18. N. Dingenouts, M. Ballauff, *Acta Polymerica* **49**, 178 (1998)